HISTORY OF ARIZONA'S

CLIFTON-MORENCI MINING DISTRICT

—a personal approach

Volume I: the Underground Days

Merry Christmas
Dad & Mom

by

Ted Cogut

and Bill Conger

Best Wishes,
Bill Conger

Library of Congress Catalog Card Number: 99-75051

ISBN 0-9675347-0-4

Published by:

Mining History
P.O. Box 1319
Thatcher, AZ 85552

CONTENTS

***About the cover picture*:** In a painting by T.L. "Stoney" Harby, cavalry troops look for ore specimens at a location that appears very much to be Eagle Creek. Stoney's wonderful painting seemed to have been made to order as a cover picture for this volume about the early days of mining in the Clifton-Morenci District for it was at the mouth of Eagle Creek that Brigadier General James Carleton's cavalry troops discovered gold "as high as 40 cents to the pan."

Stoney Harby came to Clifton in 1988 to rebuild the Wagon Wheel bar on Chase Creek Street, making it look like a bar of the wild, wild West. He also opened an art gallery opposite the Clifton Town Hall and painted many western scenes and portraits of local residents. During World War II, Stoney was a combat artist with the U.S. Marines. He later attended the Chouinard Art Institute in Los Angeles and both the Ecole de Beaux Art and Academie Julian in Paris. Stoney had one-man shows at the Gallerie de Seine and also exhibited at the Grand Salon, Salon d'Hiver and Salon d'Autome. Upon returning to the U.S., Stoney painted over 50 backgrounds for Cecil B. DeMille's epic, *The Ten Commandments.*

Stoney moved to Cottonwood, Arizona in 1991. He exhibits and receives prizes for his paintings at yearly fairs. At the Yavapai Fair he received an "Honor Award for Exceptional Merit." Stoney resides at 1600 E. Skyline, Cottonwood, Arizona 86326. He returns to Clifton each year "to show my latest paintings to dear friends."
Picture courtesy Phelps Dodge Morenci.

Acknowledgments

No book can ever be written that does not, in the writing of it, impinge in some way on the activities of others. Bill and I would like, first of all, to thank our wives, Marie Cogut and Joan Conger, for their forbearance. And also our children who are adults and live far away but were nevertheless impacted as they responded to the questions and to the changes of trip scheduling that inevitably arose as this history got under way.

At Phelps Dodge Corporation, we would like to thank Doug Yearley, Steve Whisler and Tim Snider; and at Phelps Dodge Morenci, Inc. Harry M. "Red" Conger, Bruce Richardson, Gary Jones for the historical information he provided, and Ken Malloque for the fine photos he took. And there were many others: Roberta "Bobby" Troxell who helped immeasurably with research data and historic photographs and her husband Ray; Robert Musselman, grandson of Robert Metcalfe for the information he either volunteered or steered us toward, and his daughter, Susan Norfleet, great-granddaughter of Robert Metcalfe, for the great photo she took of the portrait of her great-grandfather; the wonderful folks at the State of Arizona Historical Society; the Denver Historical Society; the State of Michigan Historical Library; Assistant Librarian Joyce Woerner and the staff of the Stair Library in Morenci, Michigan; the Carnegie Library of Pittsburgh; Julie Morton of the County Clerk's office in Montmorency County, Michigan; Mary Wolfskill of the Library of Congress; Robert K. Wright, Jr. of the Center for Military History, Department of the Army; Gene Sweeney, Western European Section, State Department; Lydia Anderson, Foreign Language Department, University of Arizona; Warren Fenzi; Don Lunt; Dave Boggess; Bill Friend; Frank Alvidrez; Barbara Myers; Glasgow M. "Cal" Callicotte; Wanda Ruedas; Jack Cuthbertson; Al Fernandez; Emma Ruiz Pando; Pete Gomez; Marceline Marietti; Frank Ruedas; Hal Ward and *The Copper Era*; Charles Spezia and all the folks at the Greenlee County Historical Society; James Colquhoun, grandson of the great innovator of the Arizona Copper Company, for the information he provided and especially for the computer disk containing the photo of the painting of his grandfather; the late Bill Epler, editor and publisher of *Pay Dirt,* who had unfailingly provided inspiration to continue the search for historical roots; Gary Dillard, editor and publisher of the *Cochise County History Magazine* for his help and reminders of our self-imposed deadline; and many many more.

Clifton-Morenci District Timeline
of certain significant events

1863 —— General Carleton's troopers find gold in Eagle Creek and copper deposits near the Gila River in western New Mexico.

1870 —— Prospectors from Silver City come to the District looking for gold. Among them are Robert Metcalfe and his brother James. They find no gold but the Metcalfes discover copper ore. Bob Metcalfe takes a sample back to Silver City.

1872 —— Mining Act of 1872 spurs interest in copper mining.

- Jim Bullard, Joe Yankie, Captain Miles Joy (sent by Captain Eber Ward), Ike Stevens, Robert Metcalfe and others begin staking claims.
- The Detroit Copper Mining Company is organized under laws of the state of Michigan and acquires the claims held by William Ryerson, Jim Bullard, Joe Yankie and John Swisshelm in the area called "Joy's Camp." Captain Eber Ward of Detroit is the principal stockholder.

1873 —— Robert Metcalfe acquires operating capital from Henry Lesinsky. Henry and his brother Charles and uncle Julius Freudenthal, who are also brought into the firm, are merchants of Las Cruces, New Mexico. The group organizes the Francisco Mining Company.

1873 —— The first copper ingot of the District, smelted from ore out of the Longfellow underground mine, is produced in an adobe furnace at Rockhouse, near the foot of the Longfellow mine incline. In the same year, the smelting site is moved to Clifton.

1874 —— Francisco Mining Company is reorganized as the Longfellow Mining Company. Bob Metcalfe and the Lesinsky group separate; Metcalfe acquires what would later be known as the Metcalf mine area and the Lesinsky group the Longfellow mine.

1875 —— William Church acquires options on all the Detroit Copper Mining Company stock.

1879 —— Henry Lesinsky completes construction of the Coronado Railroad, the first mine railroad in Arizona.

1880 —— William Church comes to Morenci, Arizona Territory (formerly called "Joy's Camp").

1880 —— Henry Lesinsky purchases the first mine locomotive in Arizona.

1881—— William Church and Dr. James Douglas, a geologist, meet with William E. Dodge, Jr. in the offices of Phelps Dodge and Company in New York City.

- Church asks Dodge for a loan with which to build a smelter.
- Dodge asks Douglas to examine Church's holdings.
- Douglas examines the adjacent Longfellow deposits and reports favorably on Church's holdings.
- Phelps Dodge advances the loan, becoming a partner in the Detroit Copper Mining Company.

1882 —— William Church finishes construction on a smelter at a site on the San Francisco River about two miles downstream from Clifton.

1882 —— The Arizona Copper Company, with headquarters in Edinburgh, Scotland, purchases the Longfellow Mining Company and Bob Metcalfe's mines.

1883 —— Church's Detroit Copper Mining Company smelter is relocated to Morenci.

1883 —— James Colquhoun comes to Clifton (from Scotland) as bookkeeper of the Arizona Copper Company. He will rise to become its president and also president of the Arizona & New Mexico Railway.

1886 —— At a site between the old town of Morenci and the Longfellow Mine, the Detroit Copper Mining Company builds the first concentrator in Arizona. In the same year, the Arizona Copper Company builds a concentrator in Clifton where Chase Creek joins the San Francisco River.

1893 —— James Colquhoun invents a leaching process that saves the Arizona Copper Company from financial ruin.

1896 —— For the first time in the copper industry, a concentrator successfully treats porphyry (low-grade sulfide) ores. It is the Arizona Copper Company concentrator at Clifton designed by James Colquhoun.

1897 —— William Church sells his interest in the Detroit Copper Mining Company to Phelps Dodge and Phelps Dodge continues the operation under the same name (Detroit Copper).

1899 —— The Shannon Copper Company is organized. It has mines above the town of Metcalf.

1902 —— Shannon Copper Company builds a smelter and concentrator at the site that will be known as "Shannon Hill," on the southeast outskirts of Clifton and the west bank of the San Francisco River.

1912 —— Combined annual copper production of all companies in the District is approximately 80 million pounds, a record that stands up to the war year 1942 when it will be about 100 million pounds.

1913 —— Detroit Copper Mining Company begins production at its smelter that has just been constructed 2.2 miles from Clifton's populated area, on the east bank of the San Francisco River. The site will be known as "Smelter Hill."

1916 —— The flotation process, a more efficient method of concentrating copper ore, is introduced into the Arizona Copper Company's No. 6 concentrator.

1917 —— The name "Detroit Copper Mining Company" is laid to rest. Its operations are now known as "Phelps Dodge Corporation, Morenci Branch."

1919 —— Shannon Copper Company sells all its holdings to the Arizona Copper Company.

1921 —— Arizona Copper Company sells all its holdings to Phelps Dodge. Phelps Dodge becomes sole operator in the District.

1926 —— The Coronado Trail is dedicated in ceremonies held at Hannagan Meadow.

1932 —— In a shutdown that will last five long years, the Great Depression forces closure of all mining operations.

Introduction

by Ted Cogut, co-author and editor

At many places in this history we will refer to the *Morenci Copper Review,* a quarterly publication of the Phelps Dodge Morenci mining operation. Begun in 1985, that popular publication was meant for people who either worked at, retired from, or just were interested in the world's foremost but little-known mining operation known as "Morenci" or, more generally, as the Clifton-Morenci Mining District. A number of the chapters in this book and also quite a few pictures first appeared in that publication. The *Morenci Copper Review* (MCR) was terminated in February 1998.

A few years earlier, I was approached with the idea that I should ask Bill Conger to join me in writing this Clifton-Morenci history. This came at a time when I still had some hope of improving my golf game at least to the point of being able to record a better score than that of my wife, Marie. It was well known around Clifton and Morenci that I did not consider that to be a trivial goal. As every golfer knows, if you are going to improve your game you have to let almost everything else go and spend more time on the course. Above all, you don't begin to write a book. But I did not hesitate a second in my decision. Having always had a rather intense interest in history, I said I would write it and I further indicated I was sure Bill would join in the effort.

That may have seemed a bit presumptuous, but I had reason to be confident. Before becoming a miner at heart, I had been an aviation weather forecaster and ballistic meteorologist. In that profession, I quickly learned to take some time to be quite certain of your facts before making a commitment. But this one was easy. I had known Bill for a long time and knew it was difficult for him to refuse any request. After I had been directed by Carl Forstrom, Phelps Dodge Morenci President, to begin putting together the *Morenci Copper Review,* I had even convinced Bill to write the history column.[1] Thereafter, his columns appeared in every succeeding issue save one, when, while he was on a well-deserved vacation, I served as his

---

[1] Gene Dunham, general superintendent, chaired the interdepartmental meeting announcing this venture into publishing. Yours truly was made editor-in-chief and David Brooks was editor until he moved to a job in the corporate offices. A number of attendees seemed to doubt it would be a success. But the effort would be continuously well received and even after the first issue was printed Richard Moolick, recently retired Phelps Dodge Corporation president, wrote a congratulatory letter regarding it. Richard T. Moolick, letter to Carl Forstrom, November 17, 1985.

substitute. In golf, it is often said: "I'd rather be lucky than good," and luck was definitely with us when Bill agreed to write the history columns.

Another lucky break was our association with the Greenlee County Historical Society, a pleasant period that one might wish could extend on indefinitely.[2] It was a place where one would become aware of the various Clifton-Morenci histories, some of them rather short, but some rather lengthy and detailed, that had been written by local writers over the years.

In the 1920s, James Colquhoun, one-time president of the Arizona Copper Company and the Arizona & New Mexico Railway, wrote two books describing the earliest years of mining in the district. I first became aware of his writings when Barbara Myers, then secretary to the Morenci mine president (then called "manager") and later, after her move to Phoenix, secretary to Douglas Yearley, CEO and Chairman of the Board of Phelps Dodge Corporation, had rescued some out-dated papers from their storage site in the mine's powder magazine that was then to be demolished to make way for a leaching operation. Among these I discovered 45 manuscript pages of one of Colquhoun's histories. It was typed on yellow paper and titled *The History of the Clifton-Morenci Mining District.* Colquhoun's books were the only histories that I then knew of that were written by an eye-witness to the days when the miner had to keep a gun within reach to protect against the Apache raids and the lawless element that would drift into town where the pickings were good and the law was what they made it. Only while doing further research for this book did I learn of another account of those early times by another important and earlier Morenci pioneer — Bob Metcalfe. Metcalfe's experiences were dictated to his daughter and published in weekly installments of the *Copper Era* in 1951-52. We are indebted to her and the *Copper Era* for the detailed account of Metcalfe's life which is presented in chapter 5.

Two other local residents had written detailed accounts of the local history — James Patton and Roberta Watt Troxell. Patton's history, which

---

[2] Among the members of the Society were several old-time residents of the area — Al Fernandez, a local historian and true gentleman; Harriet Sweeting, a lady who also knew a lot of the local history; Harvel and Stella Cosper, descendants of pioneer ranchers; Frank Alvidrez who came to Clifton from Mexico with his parents on February 14, 1912, the day Arizona became a state; June Palmer, the granddaughter of Clifton pioneer Dell Potter; Frances Powers who lent her artistic expertise to many of the Society's projects; The Honorable Allen Minker, Judge of the Superior Court whom I followed as president; Bill Conger who followed me in that office; Charles Spezia, the workhorse of the Society's museum whose grandfather had built the museum building in 1913 during Clifton's building-boom years; Don and Jo Lunt, steadfast supporters of the Society — they were just a few of the many who contributed to the preservation of the local history.

focused on the lives of Clifton's pioneer residents, was printed by the Greenlee County Chamber of Commerce during World War II.[3] In 1956, Roberta Watt Troxell wrote her well-documented history of the district as a University of Arizona Master's thesis for which she must have spent many long hours of research.[4] "Bobby" Troxell now lives in retirement in a lovely home in Tucson with her husband, Ray, who retired from the Phelps Dodge Morenci research department in 1992. She has made her research material and historical photographs available to us for which we are very grateful.

Finally, other histories had been written by authors who were not local residents. A detailed presentation of the early years of the Clifton-Morenci district had been written in 1987 by David Myrick of Santa Barbara, California.[5] An expert on railroad history, in this book Myrick also dealt with the history of the pioneer Clifton-Morenci mining companies. In 1952, Robert Cleland wrote a history of the Phelps Dodge Corporation as a whole, a work that included some Clifton-Morenci history.[6] More local history was also presented in Phyllis B. Dodge's 1987 book about the Phelps Dodge family.[7] The historical accounts already mentioned and others, a number of writings of a technical nature dealing with mining, concentrating and smelting, provided one a picture of the district in those days. As the more recent years rolled on, a number of significant events occurred; and it became clear that a history bringing events to the present might be in order.

Still, I wondered if we were the ones who should write it. After a while I began to see that Bill and I might just be likely candidates for the task. Bill had been in Morenci since the 1950s and had actually lived in the old town of Morenci. I had arrived in 1974, had never lived in old Morenci but had seen a number of the old buildings, prominent among them the old Morenci Hotel, the DC (Detroit Copper) store and the Morenci Club, that were still then standing when I arrived but would be taken down in the next few years. Bill, furthermore, had been assigned to many different positions in the Morenci operations and had, most importantly, been in labor relations during the violent 1983 strike, an important event in labor/management relations that had not yet been satisfactorily covered. And I had been lucky (or unlucky) enough to have been the point man in dealing with the frenetic

---

[3]James M. Patton, *History of Clifton,* Greenlee County Chamber of Commerce

[4].Roberta Watt Troxell, *History of Morenci, Arizona,* 1956. "Bobby" Troxell was the librarian at the Morenci High School.

[5] David F. Myrick, *Railroads of Arizona, vol.III,* Trans-Anglo Books, Glendale, California, 1987.

[6] Robert Glass Cleland, *A History of Phelps Dodge 1834-1950,* Alfred A. Knopf, 1952.

[7] Phyllis B. Dodge, *Tales of the Phelps-Dodge Family,* New York Historical Society, 1987.

rush of environmental regulation that began in the 1970s. In addition, we had both been on board during Morenci's mining construction boom of the 1980s and 1990s.

As we began to gather material for the history, we saw that we would have much more than could handily be covered in one book. So the decision to split the material into two volumes was made. This volume, Volume I, covers the history of the Clifton-Morenci district through the underground mining years. It ends roughly near the beginning of World War II. Volume II will deal with the open pit mining years.

The reader should note that the town name "Morenci" refers to two locations. The original town founded by the pioneers was swallowed up by the open pit of the Morenci Mine in the period of the 1960s to the early 1980s. Where it could otherwise be confusing, we will refer to that original settlement as "Old Morenci." The new town, whose construction began about three miles to the south in the mid-1960s, will be referred to as "Morenci." We should also explain that the abbreviation "PDMI," used in this book, is the acronym for Phelps Dodge Morenci, Inc., the name of the Morenci mining operation that was adopted in February 1986 as part of the reorganization attending the sale of 15% of the Morenci operation, with certain exclusions, to the Japanese firm Sumitomo Metal Mining of Arizona. PDMI was the last of the reorganizations of the mining operation of which much more will be said in following chapters. Briefly, the firm had been known as Phelps, Dodge & Company when, in 1881, in response to a request for a loan, it became part owner of the Detroit Copper Mining Company that was then in Morenci. Phelps, Dodge & Company then began to operate in Morenci under the name Detroit Copper Mining Company and did so even after 1897 when it became sole owner of that mining firm. In 1917, during its reorganization as Phelps Dodge Corporation, the name "Detroit Copper" was dropped and the operation became known as "Morenci Branch, Phelps Dodge Corporation," a name it retained up to 1986.

Because of our affection for the present-day descendants of the Apache warriors of Geronimo's time, the battles with their forebears is something we did not like to deal with. Geronimo's contemporaries had attacked the early prospectors and pioneer miners with lethal effect; but looked at from their side, these native Americans were desperately defending their homelands from the European-American intrusion. In reading the account of the 1864 Woolsey survey expedition I noted that Colonel Woolsey's troopers had run across farmland where corn and beans grew and had apparently helped themselves to the product of the Indian labor. Having, as a boy, attended to such crops on a Michigan farm, I could understand that that in itself might be cause enough for war. Regardless of how we felt about it, the war with the Apache Indians *was* part of the history, and we have presented it.

That "mists of time" era when the Apache warriors roamed the area, making life hazardous for the prospectors and for the miners who shortly followed them, was indeed fascinating. As I looked at the various accounts of that time, one question kept returning. Why was the settlement named "Morenci?" It was not a common place name. Where did it come from? There were different guesses, but none that seemed close to being irrefutable. Throughout the writing of this volume, it was a question that dogged me and carried me into avenues of search that later seemed almost ridiculous. That "search for Morenci" is recounted in the last chapter. In instances such as that, we elected to tell the Clifton-Morenci story in the first person.

Both Bill and I wanted to know how certain events called "history" affected everyday lives. From that came the subtitle of this volume — "a personal approach." We told the stories of the Clifton-Morenci residents who graciously granted us the time to tell us about life in the long-gone communities of Old Morenci, Metcalf and of the old Clifton days. What was it like, with the mine shut down in the Great Depression, to cope with the reality that your hope for survival lay in being able to find a few flakes of gold along the San Francisco River, Chase Creek or Gold Gulch? These poignant memories reminded me of a day in 1981 when my family and I walked through the doorway of the "Gallery in the Sun," Ted DeGrazia's studio in Tucson. The entrance had been fashioned to look like an underground mine portal. We purchased some prints. DeGrazia was in the process of signing them when he asked, as he probably routinely did with tourists, where we were from. "Morenci," I said, and with that we had the undivided attention of one of the world's best-known and beloved modern-day artists. "That's where *I'm* from," he said, "and you know what bothers me about it? It's a place I can never go back to." And of course that was because the site of Old Morenci, where Ted DeGrazia was born, had been mined away and was then a hundred or two feet in the air of the open pit mine. DeGrazia passed away in September 1982, only a short time after our little chat with him and so we were, most regrettably, not able to interview him for this book.

As a final note, appearing at the end of each chapter title are the initials "WCC" or "TLC." WCC indicates that Bill (William C. Conger) has written that chapter and TLC that I am the culprit.

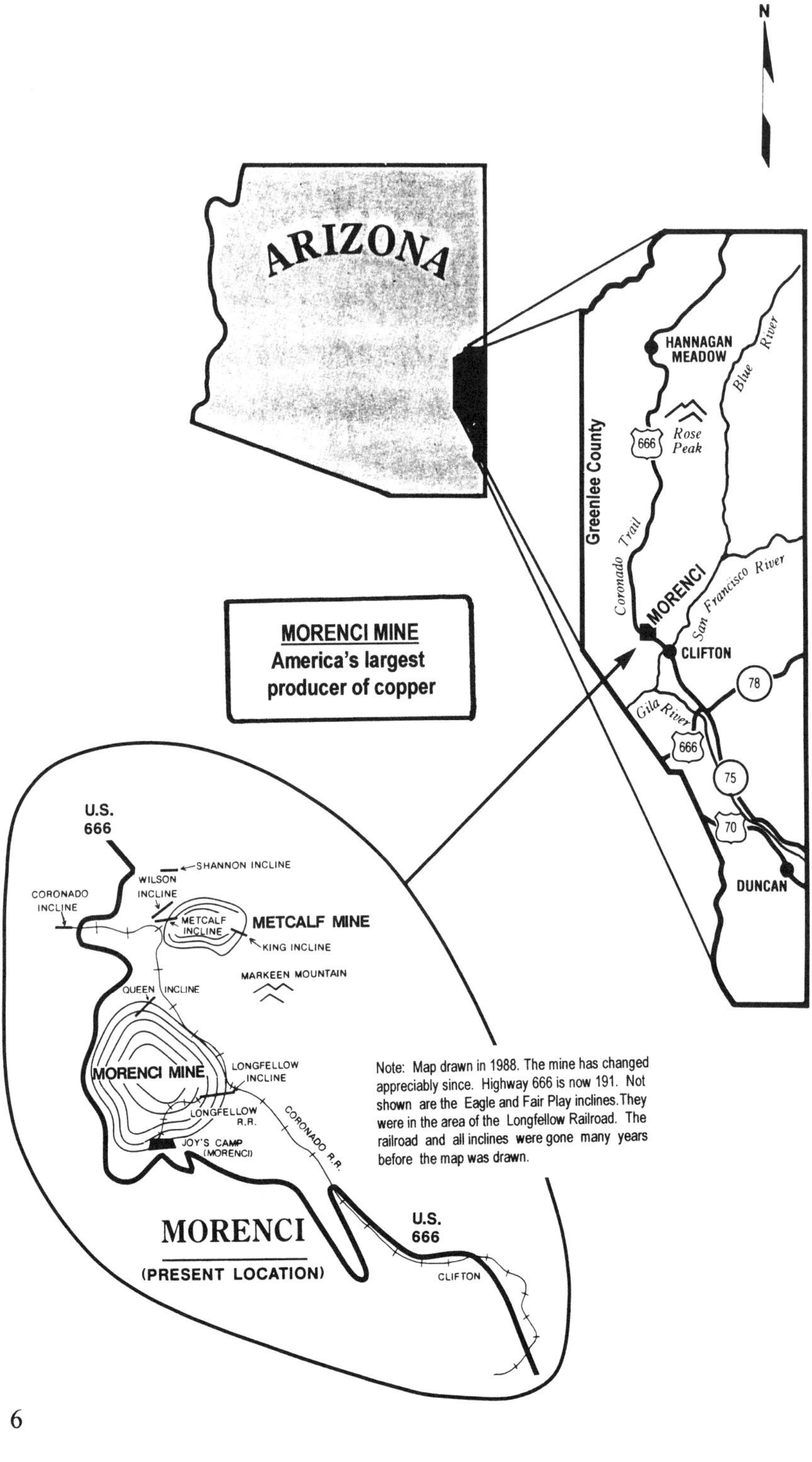
N
ARIZONA
HANNAGAN MEADOW
Blue River
Rose Peak
666
Greenlee County
Coronado Trail
MORENCI
San Francisco River
CLIFTON
78
Gila River
666
75
70
DUNCAN
MORENCI MINE
America's largest producer of copper
U.S. 666
SHANNON INCLINE
WILSON INCLINE
CORONADO INCLINE
METCALF INCLINE
METCALF MINE
KING INCLINE
MARKEEN MOUNTAIN
QUEEN INCLINE
MORENCI MINE
LONGFELLOW INCLINE
LONGFELLOW R.R.
CORONADO R.R.
JOY'S CAMP (MORENCI)
Note: Map drawn in 1988. The mine has changed appreciably since. Highway 666 is now 191. Not shown are the Eagle and Fair Play inclines. They were in the area of the Longfellow Railroad. The railroad and all inclines were gone many years before the map was drawn.
MORENCI
(PRESENT LOCATION)
U.S. 666
CLIFTON

1

A search for copper....TLC

In a book titled *The History of the Clifton-Morenci Mining District,* James Colquhoun wrote: "When Metcalf discovered the mines in the year 1870, he found stone hammers in a small trench which had been dug in the beautiful green carbonate of the Oriental Mine of the Metcalf Group. The hammers, perhaps, had been used by the prehistoric inhabitants, traces of whose encampments were still to be found in the valley of the 'Frisco; and the green ore was doubtless used as a pigment to ornament their pottery and for other decorative purposes."[1] The Metcalf Group was located a short distance north-northeast of and practically adjacent to another early mining site that would later be known as "Morenci."

More will be said in this book about Robert Metcalfe (This is his spelling of the name. The trailing "e" was dropped by others to become, simply, "Metcalf.") and more about James Colquhoun, the great innovator who came to Clifton-Morenci from Scotland to be employed as a bookkeeper. Colquhoun would save the company from financial ruin and rise to become president of the Arizona Copper Company and of the Arizona and New Mexico Railway.

Santa Rita, since 1987 the site of Phelps Dodge's Chino Mines Company, is said to be the first copper district in the Southwest to be mined by European-Americans while Ajo, a copper mining operation obtained by Phelps Dodge in 1930, is believed to be the second.[2] Like at Metcalf/Morenci, native Americans had mined the Ajo ores for pigments. The name "Ajo" derives from the Papago Indian word for paint.[3]

---

[1] James Colquhoun, *The History of the Clifton-Morenci Mining District,* London: John Murray, 1924, 32.
[2] Ira B. Joralemon, *Transactions of the American Institute of Mining Engineers, Vol. XLIX (1914),* "The Ajo Mining District," 593.
[3] Forrest R. Rickard, *Exploring, Mining, Leaching, and Concentrating of Copper Ores as related to the Development of Ajo, Arizona,* "How Ajo was Named," published by Forrest R. Rickard, Ajo, Arizona, 1.

Though it is likely that they had contact with the Papago Indians who made use of the colored copper ores, it is not known if the Indians who mined the Morenci ores had any contact with those other native Americans who mined the rich copper deposits of upper Michigan. It was most likely at that location that copper mining in search of copper as a metal was first done within what would later comprise the boundaries of the United States of America. Large-scale mining in that area would begin in the modern era in the 1840s by the European-Americans and would become an important part of the push for industrialization. Many of these miners who extracted copper near that southern shore of Lake Superior, where winters were cold and snows deep, came from the Cornwall tin mines in England. During the Civil War, copper supplied by the Michigan mines would be another of several Northern advantages.[4] But much earlier, native Americans had mined and probably made tools of Michigan copper.

In that region, unlike the Morenci area, copper was abundantly found in its metallic state. This "native copper" looked like the copper the average non-miner is familiar with. It was real red metal. Samples of copper in that form would occasionally be found in the Morenci Mine, but only enough to be a curiosity to be marveled at. Instead of being found in the true metal form, copper at Morenci would be gleaned from minerals that, to the uninitiated, would look rather like common rocks.

At first so rich in copper content as to be sent directly to a smelter to be turned into the coveted red metal, the Morenci ore later would have to be concentrated to render it suitable for smelting and would also be leached in a hydrometallurgical copper-extraction process. By contrast, in upper Michigan native Americans could simply remove metallic copper from where it lay in their mines.

An 1850 report by United States geologists described prehistoric copper mining in an area of Isle Royale, an island in Michigan's portion of Lake Superior.[5] They wrote that no metal tools had been found there but a number of stone hammers and wedges had been used to excavate the rock. The extent of that prehistoric mining activity was estimated in an 1862

---

[4] As evidence of the importance of copper to the war effort, when the Civil War began, the Confederate States of America seized the copper mines at Ducktown and nearby Copperhill, Tennessee. Copper had been discovered there in 1843. Union forces took the nearby railroad in 1863, forcing the mines to be shut down for the remainder of the war. Many years after the war mining revived. Eventually, in 1984, the Tennessee copper-mining era ended. *Mining in the Copper Basin,* Ducktown Basin Museum/Copper Basin Area Chamber of Commerce brochure.

[5] C. Harry Benedict, *Red Metal, the Calumet and Hecla Story,* University of Michigan Press, Ann Arbor, Michigan, 1952, 24.

Smithsonian Institution publication to have been the equivalent of *10,000 men working over a period of 1,000 years.*[6]

In 1991, Hal Ward, mainstay reporter for the *Copper Era*, who came to Clifton in the 1980s, showed me an ax head that had been provided to him by his brother, John A. Ward, who lived in southeastern Illinois near the Indiana border. Covered entirely with a rich green patina, to me it was obviously an ancient implement. Entrusting it temporarily to my care, Hal wanted to know if I could tell him what the ax head was made of. We removed a tiny nick and, upon being analyzed, found that it was nearly pure copper. John Ward later wrote that the ax head had been found in 1963 in a field south of Vincennes, Indiana. Could this tool have been fashioned by an early native American from native copper dug out of those ancient copper mines of what would later be known as upper Michigan? Those prehistoric miners most likely would have traded their copper, and it does seem plausible that it could have found its way to locations as far away as southern Indiana. Copper, so vital to modern man, might have, in those ancient days, been just as prized a commodity.

But let us now return to that place that would, much later, be known as Morenci, Arizona. The Morenci copper mine is the highlight attraction of any trip on Highway 191, also known as The CoronadoTrail.[7] Along that highway, as one proceeds northward into a transition zone where desert plants begin to give way to juniper and Arizona cypress, lies the Morenci Mine. About two miles in length and nearly as wide, the mine lies just north of the second townsite of Morenci, the first having been swallowed up by the open pit, the mining method adopted in 1937 to take the place of the underground workings of the earliest days. The mine is situated adjacent to an area favored by the wildlife enthusiast where deer, elk, bear, turkey and Rocky Mountain sheep are frequently seen. Even the most jaded traveler is sure to be captivated by the natural beauty of the area as well as by the majesty of the Morenci Mine itself where modern-day miners fashioned an attractive amphitheater-like structure of encircling mine benches.

Except for those native Americans who, with their stone hammers, dug out the bright green rocks in what would later be known as the Metcalf area of the Morenci Mine, Morenci's copper ores were long left untouched. The earliest European-American prospectors who descended on the area were not looking for copper — they were lured by the false expectation of finding large amounts of gold.

---

[6] C. Harry Benedict, 26.

[7] Dedication ceremonies for the newly built Coronado Trail were held at Hannagan Meadow June 19, 1926. Apache Chief Al-cha-say, Arizona Governor George W.P. Hunt and Phelps Dodge Morenci Branch Manager Joseph P. Hodgson were among the dignitaries in attendance.

The first recorded indication of Morenci-area metal ores was made in 1863 when troopers under the command of Brigadier General Carleton passed through the area on an Indian scouting expedition and discovered gold mineralization nearby in Eagle Creek. They also found copper deposits near the Gila River in western New Mexico.

The first prospectors in the Morenci-Metcalf area were not only not looking for copper, they had little interest in it. At that early date, to think that one could make a profit from copper mining must have seemed to be a fool's dream. If the copper ore were to be mined, it would have to be smelted to be made into copper metal, and there were no smelters in the area. Ores might be shipped all the way to Swansea, Wales, the world center of copper smelting, but the transportation cost would be high and probably prohibitive.

In July 1870, prospectors from Silver City, New Mexico visited the Clifton-Morenci District where they intended to pan for gold in Gold Gulch. Among the 46 men in that party, heavily armed to protect themselves from the Apaches, were Ike Stevens and Robert and James Metcalfe. While waiting for the summer rains to provide the water needed for panning in Gold Gulch, they camped at Eagle Creek.[8]

But as every Arizona resident knows, the summer monsoon rains do not always appear when expected, and that was the way it was that summer. While waiting for the rains, Robert and James Metcalfe made note of the copper mineralization in the areas that would later be called "Morenci" and "Metcalf." After waiting four weeks, the prospectors ran short of food and returned to Silver City. Bob Metcalfe took samples of the copper ore back to Silver City.[9]

In May 1872, enactment of the Federal Mining Law encouraged another party from Silver City to visit the District. Included in this party of eight were Jim Bullard and Joe Yankie (for whom a mine and concentrator would later be named). While they were staking claims, also from Silver City came Captain Miles Joy (sent by Captain Eber Ward of Detroit, Michigan) and Ike Stevens. They began doing the necessary work required for patenting the claims. This formerly isolated area might now have seemed a bit crowded, for a third party from Silver City also arrived at about the same time. They were led by Bob Metcalfe who was backed by the Lesinsky group

---

[8] Norman Carmichael and James Kiddie, *Transactions, American Institute of Mining and Metallurgical Engineers, Vol. LXX, 1924,* "Development of Mine Transportation in Clifton-Morenci District," 828; and *Copper Era,* "Reminiscences of the West," Installment No. 40 of the autobiography of Robert Metcalfe, 1952.

[9] Carmichael and Kiddie, 828; and "Reminiscences," Installment No. 40.

(merchants of Las Cruces, New Mexico). Metcalfe located claims in the area that would later become the Metcalf and Longfellow mines.[10]

Eventually, a few of the hard-working, determined pioneers would decide to extract the copper minerals on a commercial basis. The first of these organized themselves in 1873 as the Francisco Mining Company. In 1874, they reorganized as the Longfellow Mining Company. Henry Lesinsky was the principal partner. Before he brought railroading to Clifton-Morenci, Lesinsky sent the ore out on wagons pulled by bull teams with the almost unbelievable destination of Kansas City, a distance of more than one thousand miles! During this time of the Indian Wars, the trip was a perilous undertaking. Teamsters risked and sometimes lost their lives in the effort to bring the red metal to a society reaching for the more comfortable life that copper would help bring.

These pioneers were only a little ahead of their time. To be sure, copper had long been used in this country in the manufacturing of a vast number of items, such as buttons, cartridges, wire, pots and pans. In the early years of our nation, much of this copper was imported from England by a firm named "Phelps, Dodge & Company." Beginning in the 1840s, Michigan's Upper Peninsula would become an important source of the metal frequently called "Lake Copper."

By 1861, Phelps, Dodge & Company had diversified into many areas. In that year, it supplied the wire for the nation's first transcontinental telegraph line, foreshadowing the use of copper in other more far-reaching technological advances. The company had not yet become involved in copper mining, that was still some 20 years in its future.

An explosion in the use of copper would take place with Alexander Graham Bell's patent on the telephone in 1876 and Thomas Edison's invention of the first practical incandescent lamp in 1879. Edison developed an urban electrical-distribution system which, in the hands of his followers, brought light to the world. These new uses for copper brought on a sustaining and critical need for the red metal. When that happened, the pioneer Clifton-Morenci copper producers would be poised to supply the demand. To help in that effort they would bring the firm Phelps, Dodge & Company into copper mining. Through their combined efforts, which were often no less than heroic, these pioneer miners laid the foundation for what would one day become *the* world-class copper mining operation now called "Phelps Dodge Morenci, Inc."

---

[10] Carmichael and Kiddie, 828-829.

Major General James H. Carleton (1814-1873). He advised the Lincoln Administration of the mineral riches of Arizona and New Mexico, and of the gold in Eagle Creek that was as high as "forty cents to the pan."

From the collections of the Library of Congress.

2

Overview of local mining companies to the 1920s.... WCC

The Francisco Mining Company was the first company to mine copper in the Morenci-Metcalf district. The company was organized in 1873 as the Francisco Mining Company and in 1874 was renamed the Longfellow Mining Company. The rock they mined was extremely rich in copper minerals. The beautiful blues, greens, reds and blacks of the ores are undoubtedly what the first discoverers saw in the outcrops in 1870. These, along with the possibility of finding gold, brought in the prospectors, risking life and limb. The early mining companies in the district mined these rich veins and replacements. Because the ore was so high in grade, it could be charged directly into the primitive smelters. Some of the ore assayed as high as 60% copper. Ore from the Longfellow Mine averaged 20% copper over a 10-year period. After smelting, the resulting black copper [1] was hauled away as ingots in wagon trains, and later by rail, and sold.

Phelps, Dodge & Company was introduced to copper mining in 1881 when it helped the Detroit Copper Mining Company build a smelter. Organized in 1872, Detroit Copper did not begin full-scale mining operations until 1881 when William Church acquired operating capital from Phelps Dodge.

In 1897, Phelps Dodge purchased the entire Detroit Copper Company from William Church for $1,800,000.[2] Church had failed in his attempt to make the transition from treating higher-grade oxide copper ore to the lower-grade sulfide ore. The company immediately set out to solve the problem by building the West Yankie concentrator near the West Yankie underground mine, between the towns of Old Morenci and Longfellow. That concentrator, completed around the turn of the century, successfully treated

---

[1] Copper from the early Longfellow mining operations did not undergo the purification step of converting, a process that would be adopted later. James Colquhoun in his *The History of the Clifton- Morenci Mining District,* 8, 62, stated that those early smelting years produced a bullion that assayed only 88%, but the ore was so rich that as late as 1877 Longfellow ore averaged 20% copper. John Murray, London, 1924, 35, 62.

[2] *Denver Republican,* May 26, 1901, 8.

the lower-grade sulfide ore. The area of the mines and concentrator has today been mined out as the huge open pit of the Morenci Mine.

Phelps Dodge continued as the Detroit Copper Company after gaining full ownership. The mines, all underground, were located around and under Old Morenci. The larger mines were the Yankie, West Yankie, Arizona Central, Montezuma and Lone Star.

By the turn of the century, under the leadership of James Colquhoun, the great innovator, the Arizona Copper Company was successfully mining, concentrating and smelting sulfide ore. ACC ores were smelted in a smelter about six miles down Chase Creek Canyon where Chase Creek emptied into the San Francisco River in Clifton.

In 1901, the Arizona Copper Company built a small copper concentrator near the old Longfellow Mine. It only ran for a short time because the Longfellow Mine was abandoned. The mighty Longfellow had been depleted of its high-grade ore. This concentrator was located in an area later called "Dispatch Hill" in the southeast corner of the open pit of the Morenci Mine. ACC then began mining operations in its Humboldt Mine, an underground mine located on the north side of the town of Old Morenci. It was to become the largest, most productive underground mine of all those in the District.

The three major copper companies in the district in the early 1900s were the Detroit (owned and operated by Phelps, Dodge & Company since 1897), the Arizona and the Shannon. Many small copper mines such as the King, Queen, Stevens and Standard were also active.

In 1906, Arizona Copper began construction of the largest concentrator in the district — the No. 6 concentrator. It handled the lower-grade sulfide ore from the Humboldt Mine and reached a capacity of 5,000 tons of ore per day.

James Colquhoun retired as manager of Arizona Copper in 1904. At the request of the board of directors he accepted the position of president but retired again in 1907 for reasons of health and moved to Great Britain. He didn't seek out the "ol' rockin' chair" for his retirement days. Instead, he journeyed to Russia to run a copper mine in the Caucasus which he described in his book *Adventures in Red Russia.* The Arizona Copper Company continued to prosper after Colquhoun left.

Colquhoun's grandson, also named James Colquhoun, toured the Morenci Mine in 1993 and visited with Ted Cogut. In an article in the November 1993 issue of *Morenci Copper Review*, and also in a chapter of this book, Ted described that visit and presented the brief autobiography of the elder Colquhoun, a man who was important to the success of mining in Morenci and who had made important contributions to the copper mining industry as a whole. James Colquhoun was, in addition, much loved by the people of Old Morenci and Clifton.

In 1908, the old Phelps, Dodge & Company was dissolved and a new corporation was formed. The four subsidiary companies were Copper Queen at Bisbee; Moctezuma Copper Company at Nacozari, Mexico; Detroit Copper Mining Company at Morenci; and Stag Canyon Fuel Company at Dawson, New Mexico.

In 1913, The Arizona Copper Company constructed a new smelter south of Clifton. This replaced the one in the heart of Clifton which the company had operated since 1882. There were several reasons for the move. More space was needed for the depositing of slag. Slag from the old smelter had been dumped along the banks of the San Francisco River and Chase Creek. Old-timers say this was done so that the slag could in effect become a levee which would help keep the river from forming a new channel during high-water periods. Clifton has been flooded many times since the slag walls were developed, but the river has always returned to its channel. (For those who may have wondered about any environmental effects of that practice, in tests conducted under EPA-prescribed procedures it was found that the slag does not leach — it is benign in the environment.) Smoke was another reason for moving the smelter. Even with the stack at the top of the adjacent mountain, smoke could, under certain meteorological conditions, swirl down on the town.

The new smelter south of Clifton, at the site that would be called "Smelter Hill," was the largest, most modern in the district and remained so throughout the rest of the underground mining days. Up until February 1997, the site of the old smelter was easily spotted as it was marked by a 300-foot-tall stack. At that time, the stack was taken down.

In 1914, the Arizona Copper Company built a magnificent mansion on Clifton's east side for its president, Norman Carmichael. It was called "The Carmichael House." The beautiful old six-bedroom home now serves as the guest house for Phelps Dodge.

During World War I, in 1915 and 1917, bitter labor strikes against all the copper companies in the district devastated the economy. The strikes were particularly bad because the shutdowns occurred at a time when copper prices were high because of the war.

Phelps Dodge had reorganized in 1917, changing from a holding corporation to an operating corporation. All the subsidiaries were changed to branches of Phelps Dodge Corporation and the name Detroit Copper Mining Company was changed to Morenci Branch of Phelps Dodge Corporation.

The Fabulous Clay Ore Body

The most important discovery around the turn of the century was the huge Clay ore body. Its principal ore mineral was chalcocite, a copper sulfide. Up to that time almost all of the mining was of the high-grade copper carbonates, oxides, silicates, sulfates and sulfide veins. This new ore body was much lower in grade and extended onto Arizona Copper Company and Detroit Copper Mining Company (Phelps Dodge) properties. Coordinated exploration of the Clay ore body began by both companies in 1903. By 1917, the Arizona Copper Company had blocked out 47 million tons in the ore body with copper content ranging from one and one-half to two percent. From 1918 to 1920, both companies continued exploration and development.

After gaining ownership of the entire district, Phelps Dodge, in the 1920s and 30s, continued exploring the Clay ore body, resulting in the decision to mine it by open pit methods. Mining of the Clay ore body and its extensions has been carried on successfully since 1937 and has made the Morenci Mine one of the largest-producing of world copper mines.

An economic depression followed the end of World War I, plunging the price of copper from a high of 33 cents a pound during the war to 10 cents. Metcalf was badly hurt and never recovered. The town had grown since its beginnings in the 1870s to a busy, thriving mining community. It had two company stores, owned and operated by the Arizona and Shannon Copper companies, the luxury of electricity, an elementary school and a beautiful new high school that was built in 1917. The town reached a population of around 5000. It was the railhead for the Shannon and Coronado railroads.

The post-World War I depression forced closure of all the mines, which were the sole support of the town. Several attempts, largely unsuccessful, were made to run the Coronado Mine. It closed for good in 1923 and it, too, was abandoned. The citizens of Metcalf kept hoping that the mines would reopen and that their jobs and homes would be saved. But all these hopes faded as the companies began moving out, and the people began abandoning the town, dismantling their homes and hauling them away. All this happened during the 1920s and 30s.

In 1940, the beautiful new high school was sold as salvage to a scrap dealer for $1,000! By 1951, when your writer came to Old Morenci, the only evidence left of Metcalf were the walls of the old 3-B grocery store and saloon, standing forlornly on a hillside above Chase Creek (3-B stood for "Bueno, Barato y Bonito," meaning "Good, Cheap and Pretty").

By 1919, the Shannon Copper Company gave in to continual losses and sold their property to the Arizona Copper Company.

Depressed copper prices continued and by 1921 the Arizona Copper Company sold all its holdings to Phelps Dodge Corporation. Arizona Copper had been in existence 39 years — from 1882 to 1921. It was the largest company and produced the most copper of all the companies operating during the underground mining days. Under the leadership of James Colquhoun, the company survived one disaster after another, including depressions, fires, floods, mine cave-ins and Indian raids. The company survived because of Colquhoun's ability to overcome disasters through invention and innovation; but it finally succumbed to the 1920s depression, 14 years after Colquhoun had retired and left.

Phelps Dodge becomes the sole operator in the district

Phelps Dodge Corporation, Morenci Branch, then became the only operating mining company in the district. Normally, an economic depression of that magnitude would have been fatal for all the companies, including Phelps Dodge. But Phelps Dodge had enough financial reserve to survive, partly because of funds available from the old Phelps, Dodge & Company mercantile business.

The depression ended in 1922, and the copper market rebounded. A brief attempt was made to mine the Coronado Mine above Metcalf, but that underground mine was closed for good in 1923. Meanwhile, the surface had barely been scratched in the huge low-grade sulfide ore body of the Humboldt Mine which Phelps Dodge had acquired from the Arizona Copper Company. Located on the north side of Old Morenci, it became the most productive underground mine in the district. Phelps Dodge had also acquired the No. 6 concentrator in Old Morenci and the Clifton smelter, both of which had been owned by the Arizona Copper company.

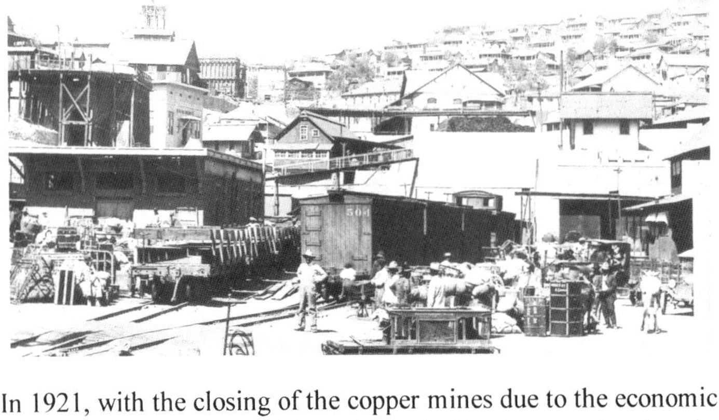

In 1921, with the closing of the copper mines due to the economic depression that followed World War I, Mexican nationals who wanted to leave Morenci and return to Mexico were given free transportation to El Paso on railroads owned by Phelps Dodge.

From the Merna Robison collection.

3

1863: General Carleton's troopers find gold in Eagle Creek and copper to supply the world....TLC

It was the letter sent to Brigadier General James H. Carleton, U.S. Volunteers, on September 19, 1863 by Surveyor General John A. Clarke, that prompted Carleton to take up his pen and write his own letter on the same general subject. Surveyor General Clarke, who, along with Carleton, was based at Santa Fe, New Mexico, had written:[1]

> Knowing that you feel a deep interest in all matters tending to the development of the material resources of this Territory, I take the liberty of communicating to you some facts relating to the newly discovered gold fields north of the Gila River, ascertained by me during my late tour of observation in that region.

On that tour, Clarke had traveled 320 miles west from Albuquerque and then 100 miles toward the southwest where he observed placer mining along the tributaries of the "Hasiampa."[2] He further described the location as being about 150 miles from Fort Mojave and 220 miles from Tucson, the two nearest military forts. It would be important to know the locations of the forts should the Tonto Apaches, who were then friendly toward the miners, have a change of heart.

Clarke estimated there were "one thousand persons of whom eight hundred are Mexicans from Sonora" at work in the area comprising Antelope and Indian creeks, tributaries of the Hasiampa, and including the lower Hasiampa area. "Between the Antelope and Indian creeks," Clarke wrote, "is the Antelope Mountain, on the summit of which is a placer of most extraordinary richness....They have taken out since the discovery a few weeks since twenty thousand dollars." On our present-day maps, we find Antelope Peak at about 20 miles north of Wickenburg, Arizona.

---

[1] Library of Congress, letter from Surveyor General John A. Clarke dated September 19, 1863 to Brigadier General James H. Charlatan, U.S. Volunteers, Major 6th Cavalry.

[2] "Hassayampa" on today's maps.

On September 20, 1863, the day after he received Surveyor General Clarke's letter, Brigadier General Carleton, Headquarters, Department of New Mexico, wrote to Salmon P. Chase, Secretary of the Treasury in the Lincoln administration. Clarke's letter had dwelled on the gold discoveries, Carleton was also interested in the gold but in other metals as well and that would include copper. Carleton wrote:

> Knowing the great interest which you feel in all matters that will increase the prosperity of the country, and more particularly at this time in all matters relating to its moneyed resources, I have ventured to write to you concerning the new Gold Fields recently discovered near the San Francisco Mountains on the 35th parallel, and between the Rio Grande and the Rio Colorado.

In covering the area from the Rio Grande, which runs north-south through Albuquerque, and the Rio Colorado, which is on the western border of Arizona and also runs north-south, Carleton was taking in a large expanse of territory. He was not relying entirely on what Surveyor General Clarke had described. His letter to Secretary Chase also included this statement which must have caused a considerable stir among those who became aware of it: "On the Prieta affluent to the Gila from the north, gold was found by my scouting parties last winter as high as 'forty cents to the pan'."

In the Greenlee County Historical Society in Clifton, there is an atlas that was printed in 1870. It had belonged to William F. Crosby, Bill Conger's grandfather. The Crosby's lived in a Victorian house with a tower that overlooked the Hudson River at Nyack, New York. Mr. Crosby passed away before Bill Conger was born, but Bill's grandmother, Carrie Crosby, lived on to the age of 92, passing away in 1952. Before she died, Grandmother Crosby gave the 1870 atlas to Bill Conger, and he and his wife, Joan, subsequently donated it to the museum.

Some of the place names shown in that atlas for the Arizona Territory have since changed. Other modern names do not appear at all. There is, for example, no mention of Phoenix — it didn't exist in 1870. Closer to home, we see that Eagle Creek, which lies only five miles west of Morenci, was then called "Rio Prieta," (dark river). This is "the Prieta" where General Carleton's troops found gold "as high as forty cents to the pan."

In describing the mineral resources for the entire Arizona-New Mexico region, Carleton compared it to California:

> From what he (Surveyor General Clarke) says and what I learn from other sources, a large region of country extending from near the head of the Gila along the southern slope of the Sierra Blanca, Sierra Mogollon (Copper Mountains),

San Francisco Mountains and thence to the Colorado, is uncommonly rich, even compared with California, in gold, silver, Cinnibar (sic) and copper."

The San Francisco Mountains are located along the Blue River about 44 air miles north-northeast of Morenci, near the New Mexico border. Carleton continued:

> You will, without the shadow of a doubt find that our country has mines of the precious metals unsurpassed in richness, number and extent by any in the world. Rich copper in quantity enough to supply the world is found at the head of the Gila. Some of this copper abounds in gold. Some is pure enough for commerce with but very little refining. The gold is pure.

The head of the Gila is in New Mexico. Though it is not at the head of the Gila, and we are not sure if Carleton knew that exact location, it is interesting to note that the Phelps Dodge copper mine at Tyrone, New Mexico, about 40 miles east of the Arizona border, lies fairly close to that stream.

Carleton sent copper and gold samples to Secretary Chase. "I send you herewith," he wrote, " a specimen of copper from near Fort West on the Gila, and two specimens of pure gold from the top of Antelope Mountain spoken of by General Clarke." *Two* locations for Ft. West are shown on the Conger atlas. They are on the Gila River and are fairly close to each other. This could be the result of a printer's error. It is also quite possible that the fort's original location had been deemed unsatisfactory for reasons such as its vulnerability to attack, its water supply, or for health considerations and a second fort constructed not far from the first. One of these is shown to be due west of Pinos Altos, New Mexico. The other is due west of the Santa Rita copper mine (purchased by Phelps Dodge in 1987) and quite close to the Phelps Dodge Tyrone, New Mexico copper mine (which did not exist when the map was printed).

Carleton then asked for troops that could be used to protect prospectors from the Apaches "so that prospecting parties can explore the country and not be in fear all the time of being murdered..." Considering the need for soldiers to fight in the East, this was probably a futile request.

In September of 1863, the time that Carleton wrote his letter, the Civil War had been dragging on for two and a half years, and, in spite of the recent Union victories in July at Gettysburg and Vicksburg, the mood in Washington had to be somewhat dreary. Perhaps that was part of the reason why Carleton included the following request with the copper and gold specimens he sent to Secretary Chase: "If it be not improper please give the largest piece of the gold to Mr. Lincoln. It will gratify him to know that Providence is blessing our country even though it chastiseth."

The piece of gold that was to be given to President Lincoln had most likely been found by troopers under the command of Captain Pishon who was sent by Carleton to escort Surveyor-General Clarke to the gold fields "near where the City of Prescott, Arizona now stands."[3] Captain Pishon's instructions were contained in this letter:

Headquarters Department of New Mexico,
Santa Fe, N.M., June 22, 1863

Capt. Nathaniel J. Pishon, *Company D, First California Cavalry, Fort Craig, N.M.:*

CAPTAIN: I send a map of New Mexico, on which I desire that you will trace your route to and from the new gold fields, in obedience to orders to go as an escort to Surveyor-General Clarke.

Have great care taken of your animals. When you arrive at the new diggings, I want each of your men to prospect and wash (pan for gold), and I want you to report the exact time they severally work, and the amount of gold each one obtains in return for his labor during that time. Much reliance will be placed upon these statistics. The people must not be deceived, nor inveigled into that distant country without knowing well what they may expect to find.

If the country is as rich as represented — and of this I have no doubt — there will, on your return, be a revolution in matters here which no man now can ever dream of. I have written to the authorities at Washington that if the country is as rich as reported,on your return I shall send two companies of California troops to establish a post in the heart of the gold region. Your company may, perhaps, be one of them, so you will have an eye to the best location of a post for one company of infantry and one of cavalry. In returning by the Whipple route to Albuquerque, mark the country well for the whole way from the gold region. Take your best men with you, and things to wash with. Send me a few specimens for the War Department on your return. Wishing you good fortune,

I am, Captain, very respectfully your obedient servant,
JAMES H. CARLETON,
Brigadier-General Commanding[4]

The troopers who had made the discovery of gold that was "as high as forty cents to the pan," were in all likelihood those of Company A, First Cavalry, California Volunteers, Captain McCleave commanding. A document titled *Records of California Men in the War of the Rebellion* indicates that on January 27, 1863, Company A "left Fort McLean for

---

[3] Brig. Gen. Richard H. Orton, *Records of California Men in the War of the Rebellion,1861-1867*, Sacramento State Office, J.D. Young, Supt. State Printing, 1890, 72. (In 1866, General Orton had been a Captain in Company M and had been involved in skirmishes with the Apaches along the border of New Mexico and Mexico.)

[4] Brig. Gen. Orton, 72.

headwaters of Gila and Francisco Rivers, in accordance with Special Orders No. 5, District of Arizona. Returned to Fort McLean February fifth." So it is likely that Company A found the gold in Eagle Creek around the end of January or early February of 1863.

Through a query to the U.S. Army's Center for Military History, I learned that Carleton's service had been distinguished enough to have him recommended for the Army's Fort Leavenworth Hall of Fame. I thought this might have been due to his great discovery of the mineralization in the Morenci and western New Mexico areas, but there was no mention of that in the biographical sketch that the Army sent me. His other accomplishments, and there were many, were the ones that had merited that recommendation. The biography was eight pages long. For those of our readers who would like to know more about the man who first brought the attention of the nation to the Morenci area, here are some of the more memorable aspects of his life:

Carleton was born in Lubec, Maine on December 27, 1814. He first wanted to pursue a literary career and corresponded with Charles Dickens for advice and encouragement. It was this literary bent that would serve him well in the military career he was finally to adopt.

The Maine boundary controversy with England led Carleton into the Maine militia and from there, on October 18, 1839, on to an appointment in the U.S. Army as a second lieutenant in the First Dragoons.

In his Army assignments, you see a reflection of the history of the times.[5] One of these was to remove government property from the dismantled Fort Croghan near Council Bluffs, Missouri. He built mackinaw boats to float the government property down the Mississippi and while doing that he checked river boats for liquor which the U.S. Intercourse Law of 1834 forbade. In the course of these checks, he met John James Audubon who gave him two of his animal paintings.

At Fort Leavenworth, he was assigned to two large expeditions. On the first of these, he went to the Pawnee Indian villages in Nebraska in an attempt to improve the government's relations with the Pawnees and to prevent the frequent tribal wars between the Pawnees and the Sioux. On the second, he left to obtain information about Indian tribes along the Oregon Trail and about Oregon. He served as one of the secretaries at the various Indian Councils on both expeditions and later published his valuable notes about the Indians he encountered. The notes were titled "The First Log

[5] Department of the Army, The Center for Military History, biography of Brevet Major General James Henry Carleton, Washington, D.C.

Book" and "The Second Log Book" and were published in New York from 1844 to 1846 in the *Spirit of the Times.*

During the Mexican War, on May 30, 1846 Carleton crossed the Rio Grande into Mexico with General Wool's forces. Wool received a summons to come to the aid of General Taylor at Saltillo and arrived in time to help defeat the Mexicans at the battle of Buena Vista. Once again, Carleton documented what he saw and later wrote *The Battle of Buena Vista, with the Operations of the "Army of Occupation" for One Month.* It is the only written eye-witness account of the battle.

In one of his next assignments, at Fort Leavenworth, Carleton, now a major, compiled a "Table of Distances" which was published in July 1850 in *Stryker's American Register and Magazine.* These statistics were made for the benefit of the emigrants who were leaving the Fort Leavenworth area to settle the West.

In 1857, following duty in New Mexico to subdue the warring Indian tribes after which the Jicarilla and Mescalero Apaches, the Navajos and the Utahs all came to terms with New Mexico's civilian authorities, Carleton and his dragoons patrolled the route across the Mojave Desert and built redoubts to protect the settlers along the Salt Lake route.

We now enter the Civil War period and Carleton is promoted to colonel. In October 1861 he is given command of the First Regiment of California Volunteers followed by command of the District of Southern California, and that was followed by command of the First California Volunteers, the First Battalion of Cavalry and a battery of artillery.

While serving in that last command, Carleton received orders to reopen the southern mail route and recapture from the Confederates Forts Thorn, Buchanan, Fillmore and Bliss. By August 1862 he had accomplished that mission and captured, in addition, Forts Quitman and Davis. This would be important to the people of Texas, New Mexico and Arizona. The secession of New Mexico and Arizona and Confederate control of the Southwest had thereby been blocked.

On August 26, 1862, in recognition of that service in the Southwest, Carleton was given command of the Department of New Mexico. He kept a watchful eye on Confederate activity to ensure they did not infiltrate the area and also focused on the betterment of the Indians living in the area of his command.

Carleton established a reservation for the Navajos on the Pecos River which contained an agricultural school and experimental farm. In the summer of 1864, he established the first Indian school in New Mexico. Through hard work and kind treatment, in two years he achieved with the Navajos what took 20 years of warfare to achieve with the Apaches.

The end of the Civil War brought Carleton a promotion to two-star rank — he was then Brevet Major General of Volunteers.

James Henry Carleton died while stationed in San Antonio on January 7, 1873 at the age of 58. All but eight of his 33 years in the Army had been served in the Southwest. His many accomplishments were noted in the biography recommending him for inclusion in the Fort Leavenworth Hall of Fame, the document from which this information was taken. But nothing was mentioned about what we think was a most notable achievement — he had been the first to make known to the nation the important mineral resources of the Southwest and, of particular significnce to us, of the general Morenci, Arizona area.

1870 Atlas of the Arizona Territory. *Courtesy the Greenlee County Historical Society.*

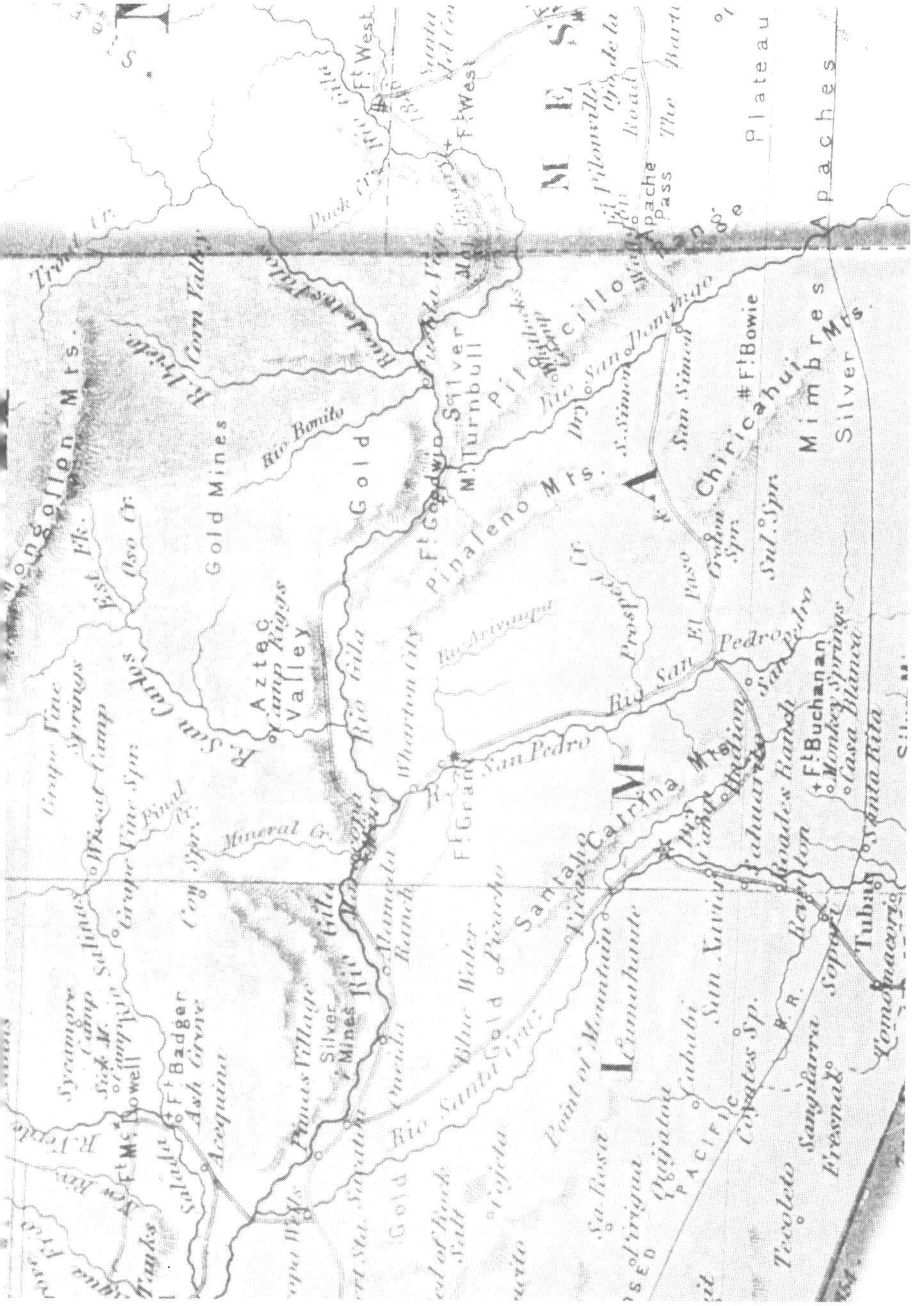

1870 Atlas of the Arizona Territory. *Courtesy the Greenlee County Historical Society.*

Salmon P. Chase, Secretary of the Treasury in President Lincoln's administration, received the letter written by General Carleton on July 20, 1863 which advised of the mineral riches to be found in the Southwest. *From the collections of the Library of Congress.*

4

1864: Woolsey finds gold in Pinal Creek and almost finds Morenci's copper....TLC

The year after Brigadier General Carleton's troops found gold "as high as forty cents to the pan" in Eagle Creek, Colonel Woolsey, with base camp on a ranch at Agua Fria, headed an expedition eastward to the Salt River area and then on to the Gila River. He provided a detailed account of his expedition to Governor John N. Goodwin of the Arizona Territory and had that account printed in an issue of *Arizona Miner* whose flag proclaimed: **"The Gold of that Land is Good."**[1]

No doubt being familiar with the great gold discoveries of the Antelope Mountain area, Woolsey had hopes of making another such discovery farther eastward in the Arizona Territory.[2]

Names on the maps at the time of Woolsey's expedition were not all the same as they are now which probably has resulted in some confusion about his trip. Some who read Woolsey's report seem to have assumed that he went to the San Francisco River, that great stream that runs through Clifton. But the San Francisco he mentions is really the Verde River that flows southward to join the Salt close to Phoenix. On our 1870 map, the present Verde River was also called the "San Francisco."

Confusion also probably resulted from his mention of the Black River which is not the Black we are familiar with, the one that flows along in an east-west direction 37 miles north of Morenci. The Black he mentions is what was then also called "Rio Prieta" and later changed to "Eagle Creek." In yet another version of the name, in one of his books written in the 1920s on the history of Clifton-Morenci, James Colquhoun calls that stream "Eagle *River*."

---

[1] *Arizona Miner*, September 24, 1864.

[2] The Arizona Territory was established in 1863. Its first capital was set up at Prescott in 1865.

One of the campsites occupied by the Woolsey party was at the top of Pinal Mountain about eight miles south of Globe where they took compass readings to prominent distant peaks including Picacho, that prominent landmark looked-for by present-day travelers as they drive along Interstate 10 between Tucson and Phoenix. When Woolsey's troopers reached the headwaters of Pinal Creek, they found gold but not in paying quantities. They then went on to Fort Goodwin.

Woolsey made several encampments at Fort Goodwin, (a site probably named after the Arizona territorial governor, John N. Goodwin) which was on the Gila River. This is also a place that later had its name changed. It was abandoned and later rebuilt at a nearby location as Fort Thomas and is now the non-military community of Fort Thomas that lies 20 miles northwest of Safford on U.S. Highway 70 that leads toward Phoenix.

In June 1946, Margherita V. Scully of Atlanta, Georgia who, as a child, had lived at the military Fort Thomas when it was built, wrote a letter that provided a vivid picture of frontier life in this part of southeastern Arizona. Roberta Watt Troxell was kind enough to give us a copy of the letter. Here is an excerpt: [3]

> My father, the late General James Wall Scully, U.S. Army, was ordered by the War Dept. to construct a four-troop cavalry post on the banks of the Gila River. There had been an infantry post nearby called 'Camp Goodwin,' but this had been abandoned on account of the prevalence of malaria. The site of the new post was practically a desert of sand. My father, then captain (and brevet colonel) named the new post Fort Thomas, after General George H. Thomas who commanded the army of the Cumberland in the Civil War. My father was aide to General Thomas and for that reason gave it his name.

We might think of how life in our area has changed since the 1880s as we go past Fort Thomas on the long drive to Phoenix. Margherita Scully further described it:

> Fort Thomas was built in 1881. My father had complete charge of the building, and even drew the plans. All the woodwork was California redwood. My mother would not have the woodwork in our quarters painted white as the plans called for, but had the beautiful redwood oiled and waxed.
>
> Our house was a large raised cottage with a wide hall and three large rooms on either side and a twelve-feet-wide porch across the front of the house. Every room had two or three large windows, and the doors had transoms to admit all the air possible. Our Chinese servants would hang army blankets wet with water on

---

[3] Margherita V. Scully letter dated June 10, 1946. This 3 1/2 page letter was provided to me by Roberta Watt Troxell. It gives a good description of what life was like during the frontier days. There is a fine equestrian statue of General Thomas in Washington, D.C.

the outside of the dining room windows so it was always cool in the rooms on that side of the house.

Our drinking water was boiled and hung in a white canvas bag by chains on the back porch where it swayed back and forth cooling the water and a small dish of limes was kept near to add to the water to counteract the alkali, only drinking water was from the Gila River.

I have heard my mother say they had to put pans of water with carbolic acid added under the bed legs to keep out the centipedes, tarantulas and scorpions.

In August 1883, the Scullys packed their belongings for a trip to Fort Adams, Rhode Island, the post to which Captain Scully was transferred. Margherita described the way a family like hers would have traveled in those days:

Several stage coaches had been robbed by so called "cowboys" and several bands of Indians were on the war path. So we were given twenty cavalry soldiers as an escort to Bowie Station.

We rode in an Army ambulance drawn by eight mules. The leaders were two tiny Mexican mules who could run like the wind, they had long bead earrings in their ears. They were used as leaders to keep the big heavy mules from balking. The soldier driver had a bucket of small stones to throw at the leaders when he wanted them to go faster....

Margherita Scully noted that their cavalry escort rode on each side of their stage coach and kept some distance behind to keep dust from the coach. One of their first stops was at Safford, where, as we now do, they found the hospitality warm.

And now, to return to the Woolsey expedition: From Fort Goodwin, Colonel Woolsey's group went to the stream then called Rio Bonito (Bonita Creek), traveled upstream about 12 miles and then turned east to the Black River (this was the Rio Prieta that later was renamed "Eagle Creek"). Woolsey described the stream:

The Black River is a good sized mountain stream, heading in a lava mountain about 20 miles south of the Mogollon mountains (Mogollon Rim), and runs in a southerly direction about 60 miles, emptying into the Gila about 35 miles above Fort Goodwin. I think we struck the Black River 15 miles above its mouth.

Woolsey closed his report with a general appraisal of the land they had surveyed:

The whole country through which we have passed is covered with excellent grass, water is plenty for all ordinary purposes. In many places beautiful little

> valleys invite the farmer and ranchero to follow the occupation of their choice. We never found gold in paying quantities, and yet I cannot help thinking that there is in that part of the Territory great mineral wealth.

Colonel Woolsey probably never realized how accurate his guess was. The year before, General Carleton had advised Washington of the gold his troopers had found at the mouth of Eagle Creek and of copper found in western New Mexico. This had brought attention to the Morenci area, but the great Morenci ore body itself escaped detection. Now, Colonel Woolsey had traveled eastward with his expedition and had reached Eagle Creek about 15 miles north of its juncture with the Gila. If he had gone a little less than 10 miles farther, in a southeasterly direction, he could easily have spotted the brightly colored copper oxide minerals of the Morenci district. That discovery was to be left for others to make.

Eagle Creek, a target of prospectors in the 1860s, in a 1995 photo by Ken Malloque. Cour*tesy Phelps Dodge Corporation.*

The Black River near the pumping plant that supplies water to Morenci and Clifton. This stream runs in a general east to west direction. The river takes a southward dip at this point and is only 38 miles northwest of Morenci. In early Territorial days, Eagle Creek, a north to south flowing stream located to the south, was also called "Black River." *Photo by Ted Cogut, 1990.*

5

1870: Bob Metcalfe knew copper would make Morenci world-renown....TLC

"It was my original intention to go to California, but when we arrived in El Paso, we learned that new discoveries of silver in great abundance had been made near Pinos Altos, New Mexico, so we decided to have a look at these new discoveries. When we arrived at the Mimbres River I found the excitement intense." This was Robert Baylor Metcalfe's recollection of his arrival in the Southwest in 1868, an event which, after a couple of year's prospecting in the Silver City area, would ultimately lead him to the rich copper ores of Morenci. In 1905, Metcalfe related these and many other events to his daughter, Maud Metcalfe James, during the dictation to her of his autobiography.[1]

One day in the summer of 1996, Susan Norfleet, the great-granddaughter of Bob Metcalfe, made a visit to the Morenci Mine. I missed seeing her by only a few hours. This was a big disappointment as I had been looking for background information on Bob Metcalfe, who, in those early days of the western frontier, probably was the first to believe it would be copper and not gold that Morenci would eventually become noted for.[2]

In a sense, Metcalfe would only rediscover an ore body that had first been brought to some prominence by General Carleton of the Army's New Mexico Department who, as we saw in Chapter 3, had written of it in a letter to President Lincoln's Secretary of the Treasury, Salmon Chase. But Carleton, though he made note of the other metals, was primarily interested in the evidences of gold of which, as it turned out, there would only be small amounts in the Clifton/Morenci district. Metcalfe, however, saw the district's future in a market that would some day be clamoring for copper.

A few days after Ms. Norfleet's visit to Morenci, I tried to call her at home in Richmond, Virginia and was put in touch with Robert Musselman,

---

[1]. "Reminiscences of the West," the autobiography of Robert Metcalfe as published in weekly installments of the *Copper Era*, 1951-1952

[2] Linda Stacy, Administrative Assistant at Phelps Dodge Morenci, called me on the succeeding day to tell me of Susan Norfleet's visit.

her father, who was a grandson of Bob Metcalfe. Mr. Musselman, an attorney in Charlottesville, Virginia, had himself, in the 1960s, visited the Morenci Mine with his wife and his mother who was born in Silver City, New Mexico.

In a number of phone conversations Mr. Musselman provided a wealth of information including the fact that the Musselman family had a wonderful portrait of Bob Metcalfe. Mr. Musselman later sent me a photograph of that portrait that had been expertly taken by his daughter, Susan Norfleet, who had made that visit to the Morenci Mine in 1996. Robert Musselman also told me that Bob Metcalfe had dictated an autobiography to his daughter, Maud Metcalfe James, Mr. Musselman's aunt; and it had been printed in a long run of weekly installments in Clifton's *Copper Era*. The articles ran from May 25, 1951 to March 21, 1952. I knew the *Era* had lost all its past issues in the great San Francisco River flood of 1983, so I hurried over to the Arizona Historical Society library (in Tucson), a wonderful place, by the way, that is well worth the visit of anyone interested in western history. The library had those past issues. I looked up that Metcalfe autobiography.

By anyone's measure, Bob Metcalfe had led a most interesting life. Born in Kentucky in 1825, after a somewhat sketchy rural schooling he passed the examination to enter West Point as a cadet in 1842. Robert Musselman had corresponded with West Point regarding his grandfather and found that he had been admitted to the academy from Missouri on July 1, 1844 and had left there January 7, 1845. His letters of recommendation had indicated that he could also obtain a recommendation from the governor of Kentucky.[3] Metcalfe, in his autobiography, attributed his early withdrawal from the academy to the fact that his farm-boy type education had ill-prepared him. "I was so handicapped," he said, "by my lack of elementary mathematics that I found it impossible to acquit myself with credit."

Metcalfe then went to Missouri where he worked for a year on a farm and then left for Texas to join the Army for the invasion of Mexico. Along the way, he stopped in Natchez, Mississippi where he met an uncle, Dr. James Metcalfe, who offered to pay his way at college. Taking advantage of that offer, Bob Metcalfe then spent two academically productive years at Jefferson College, Washington, Missouri. Another uncle, Dr. Volney Metcalfe, then offered to have him study medicine, apparently in an apprenticeship.[4] In that pursuit, he studied Latin, medicine and history for a year. Then, in 1849, news of the California gold rush reached him and he was off to join the stampede.

---

[3] Robert Musselman, phone conversation, December 28, 1996.

[4] Robert Musselman, phone conversation, December 18, 1996.

If you were to go overland to California in those days, you were telling yourself that you had agreed to accept all the risks of travel through a combat zone. The Indians, in particular the Comanches of Texas and Oklahoma and the Apaches in what would later be the states of New Mexico and Arizona, were desperately defending their homelands from the European-American encroachment. At one point in Metcalfe's journey through Comanche territory:

> The Indians placed two scalps on poles in our line of travel so we would find them. One had long red hair about two feet long and the other was short and black, streaked with grey. They were stretched over willow hoops about five inches in diameter and resembled the head of drums.[5]

But not all Indians were hostile. On another occasion, Metcalfe found himself alone outside a fort in a Mexican village along the Rio Grande when a war party of Comanches, 70 or 80 strong, rode up. After identifying him as "Mericano," Metcalfe said the Indians did not molest him in any way and demonstrated:

> Some of the greatest feats of horsemanship I ever witnessed. They charged up and down the streets at a fearful rate, and at a given signal the horses would all wheel into a circle and sit down on their haunches like dogs, with the Indians' heads along side the horses' necks.

Then on to gold mining in California and Oregon with several hair-raising adventures along the way, involving the lawless element as well as the Indian population. While in Oregon, he was made Indian agent and encountered even more hair-raising adventures, finally resigning to return east where Civil War would soon erupt.

While not in favor of secession (he would have preferred that the South stay in the union and declare itself the true U.S. government), Metcalfe nevertheless joined Terry's Texas Rangers and enlisted as a private soldier in the Confederate Army. He served throughout the four year's of war. "I served the Confederacy faithfully," he said, "and was with the last regiment to surrender and am glad of it."[6] After the war, Metcalfe turned to farming in Texas, married, and began his first family. His wife and two children, a boy and a girl, all died from consumption within four months.[7] This tragedy drove him from the farm and led him, by way of the silver deposits of the Silver City area, to his last great enterprise — the discovery of the Morenci copper ores.

---

[5] "Reminiscences," Installment No. 7.

[6] "Reminiscences," Installment No. 38.

[7] "Reminiscences," Installment No. 39.

In 1870, Robert Metcalfe and his brother James left Silver City for a brief prospecting trip to the area that would later be called "Morenci." There, they saw the rich ores that would later be part of the Longfellow and Metcalf mines.[8] In recalling that great discovery, Bob Metcalfe said:

> I took some samples of the ore with me and told the others that all this copper would one day be valuable.
>
> I was laughed at and told that I would never live to see the day when a pound of copper could be taken out from where we were for less than double what it was worth. The nearest railroad at that time was twelve hundred miles away in Kansas.[9]

In 1872, Metcalfe returned to the Morenci area with his brother, James, and another man to locate the copper mines. They discovered the Longfellow, Little Annie, Fair Play, Metcalf, Clay, Oriental and the White Hawk mines.

Maud Metcalfe James referred to that discovery in this excerpt from a letter she wrote to Roberta Troxell:

> The following is copied from an article in the *Copper Era* of Oct. 24th 1901. Written by my cousin Charles Metcalfe Shannon, who came west to join his uncle (my father), in 1871:
>
> "In 1870 a party of prospectors from Silver City, N.M. sojourned in the Clifton-Morenci district and took up claims. Copper was not much considered in those days however and little was done toward calling public attention to the district. In 1872 some of the locators returned and made good their holdings of what are now some of the most valuable mines of the district. Among these Robert B. Metcalfe proved to be the moving spirit, and he is entitled to rank as the real discoverer of the district."[10]

To attempt to mine during those days in the Clifton-Morenci area was to risk being attacked by the Apache Indians who were then actively trying to oust the European-Americans from their homelands. In another letter about her father and his associates, Maud Metcalfe James wrote: "During the first years they were in Arizona they had many attacks by the Apache Indians, and nearly lost their lives many times."[11] And mining also would require a certain amount of capital.

---

[8] In the later naming of the townsite and the mine of that name, the trailing "e" in "Metcalfe" was dropped.

[9] "Reminiscences," Installment No. 40.

[10] Letter from Maud Metcalfe James to Roberta Watt Troxell, date unknown.

[11] Letter from Maud Metcalfe James to Roberta Watt Troxell, dated October 7, 1954.

To acquire operating capital, Metcalfe took in Henry Lesinsky as a partner in his Longfellow Mine. James Colquhoun described the arrangement as one that would lead to some controversy but would end years later in a friendly reconciliation.[12] Lesinsky would bring in as his partners his brother Charles Lesinsky and his uncle Julius Freudenthal who were merchants in Las Cruces, New Mexico Territory. The firm erected a crude furnace and began to smelt copper in 1873 though there would be a number of trying years before it would enjoy a positive cash flow.

The Longfellow Mine, the big early copper producer of Morenci, was named by Bob Metcalfe. In 1954, Maud Metcalfe James wrote: "I don't know why Father named the mine Longfellow except the fact that he liked the name."[13] But like other familiar Arizona mining place names such as the "Bluebird" and the "Christmas" mines, it did seem to follow the miner's attempt to influence fate through optimistic suggestion. Henry Wadsworth Longfellow, who died in 1882, was a most beloved American poet of the 19th century. Poems such as Longfellow's "The Village Blacksmith" which extolled the perseverance and indomitable spirit of the self-reliant man would surely have appealed to Metcalfe.

In 1882, Metcalfe sold his remaining mines to the Arizona Copper Company, a firm with headquarters in Edinburgh, Scotland and returned to Kentucky, the land of his birth. He had married a Kentucky girl in 1877. They had a son, Robert, and two daughters: Maud Metcalfe James, who transcribed his autobiography, and Mrs. J.F. Musselman.

Bob Metcalfe said he built the first house in Clifton and named Clifton as well as Silver City, New Mexico and Globe, Arizona. With the trailing "e" dropped from it, his name lived on in the once thriving but long-since-gone town of Metcalf and continues to this day in the names of the mine and concentrator we call "Metcalf." He was a nephew of the man who endowed Baylor College. The late Elizabeth Shannon of Clifton, who was well known in Clifton and Morenci as well as in the state as a whole, was his great niece.

Bob (Robert Baylor) Metcalfe died in Lexington, Kentucky in December 1905 at the age of 80,[14] a man who sought out and faced the foremost challenges of his time and, in the course of that pursuit, left an indelible mark on the Clifton-Morenci mining district.

---

[12] James Colquhoun, *Early History of the Clifton-Morenci District,* London and Beccles, William Clowes and Sons Limited, c1920, 40, 79-80. In her letter to Roberta Watt Troxell of October 7, 1954 about her father, Maud Metcalfe James concluded: "That is all passed and over & better not remembered."

[13] Maud Metcalfe James letter, October 7, 1954.

[14] Robert Musselman gave this date in a phone conversation December 21, 1996. He said it was not the same as that in the "Reminiscences."

Robert Baylor Metcalfe (1825-1905) discoverer of the Morenci-Metcalf copper mines.
Courtesy Robert Musselman.

The earliest prospectors depended on the trusty burro to do almost all of the hauling. *Courtesy Phelps Dodge Corporation.*

Before the railroads, horse-drawn shipments were common. These teams are posing on a curve near Metcalf.
Courtesy Phelps Dodge Corporation.

6

Underground mining was different....WCC

From the beginning and up until 1932, when mining ceased due to the Great Depression, all mining in the Morenci-Metcalf district was conducted underground.

Mining in the earliest days — the 1870s and 1880s — could be simply described as: get in, blast, take what ore you can and get out before the whole stope caves in. Ores were mined from gigantic caverns. Pillars of ore were left at intervals to support the back, or top, but the miners would keep skimming ore from the pillars until they looked more like ornamental bedposts.[1] As a result, disastrous cave-ins could devastate the mine and threaten life and limb.

At that time, holes for the dynamite stick were hand-drilled in the rock by means of a drill steel. The miner would hold a drill steel in one hand and a hammer in the other. He would twist the drill as he hit it with the hammer. Or, if he had a great deal of confidence in the accuracy of his work mate, he could hold the drill steel while the other pounded it with a sledge hammer. In any case, it was a painstakingly slow and tough way to drill a hole.

The next step was to load the hole with explosive. Then the fuse was lit. At that point, someone yelled "Fire in the hole!" Everyone ran and devil take the hindmost! For big blasts, the miner would hold a fistful of fuses and rub fire on their ends so they were all lit simultaneously. He would have time to find cover — unless one of the fuses was a "runner," a fuse that burned much faster than the others.

---

[1] James Colquhoun, *TheHistory of the Clifton-Morenci Mining District,* John Murray, London, 1924, 12. "Timber was not used to support the ground, but pillars of ore were left at intervals. I never yet met the miner who could leave rich pillars of ore alone...."

Bill Friend is a primary blasting powder man and Dave Boggess is a mining engineer.[2] They have had many years of experience in underground mining, both at Phelps Dodge's Copper Queen Mine at Bisbee, Arizona and its Dos Pobres Mine at Safford, Arizona. They gave us this information about the "good old days" of underground mining.

At the Phelps Dodge Copper Queen Mine in Bisbee, fuses of different length were held on a split stick and all were lit at once with a spitter, similar to a fourth of July sparkler.

As time went on, improved methods were developed. Drills were run with compressed air and water. These were the "water liner" or the "jack leg" drill. The liner drill takes two men to move, whereas the jack leg is operated by one strong and experienced miner.

Later, drills were mounted on arms called booms which in turn were mounted on carriers on rail. Some carriers had two booms each and some were on rubber tires as was done in modern times at the Phelps Dodge Dos Pobres Mine in Safford.[3]

Holes in the face of the drift were drilled in specific order. Four "top" holes across the top, four "breast" holes next, four "middle" holes, four "skimmer," and at the bottom, four "lifter" holes. Three "burn" holes were drilled in the middle of the face. The holes were then shot in sequence, with the lifters going off last. Different length fuses were used, which were attached to a stick of powder into which a blasting cap had been inserted.

Today, electric blasting is practiced in some underground mines, but care must be taken that other electric devices, such as radios, do not set the blast off prematurely.

The Joy shaft on the snowy day of December 21, 1937.
Courtesy Phelps Dodge Corporation.

---

[2] From conversations held in 1994. Both Bill Friend and Dave Boggess began working in the Morenci operations in 1984.
[3] Development of limited operations at Dos Pobres began in the early 1970s. The mine was closed in the 1980s due to low copper prices.

Underground drifting—drilling holes for blasting. *Courtesy Phelps Dodge Corporation.*

Early miners strike poses for the photographer at a mine portal near Morenci. *Courtesy Phelps Dodge Corporation.*

7

1881: Phelps Dodge joins Detroit Copper Mining Company of Morenci....TLC

"Church's entire establishment," wrote Dr. James Douglas on his first trip to Morenci, "consisted of a tent and a team of horses with which he had driven down originally from his home in Colorado."[1] This was in the year 1881 and a tent and a team of horses constituted the immediately visible assets of the operation that eventually would become Phelps Dodge Morenci, Inc., the largest producer of copper of any mine in North America. But Church had more than just the tent and horses. He had obtained control of Detroit Copper Mining Company claims.

In 1872, the Detroit Copper Mining Company had been incorporated under Michigan laws, and Captain Ward, the principal stockholder, had sent Captain Miles Joy to that place in the Arizona Territory that would later be named "Morenci." At the time of Joy's residency, the site would have been referred to as "Camp Joy."[2] Miles Joy was there to survey the claims. Ward was the first in that direct, long line of extraordinary men who would be involved in the mining enterprise that would eventually become the flagship of the Phelps Dodge Corporation.

Captain Eber Brock Ward was born in Canada, in Lincoln County of the province of Ontario, on Christmas day 1811. By the time Eber was 10, the family had relocated to Detroit, Michigan. He grew up to become Michigan's first millionaire to have acquired his fortune through manufacturing.[3] His views on national financial matters "gave him a national prominence, and his name had more than once been mentioned in connection with the Secretaryship of the Treasury."[4]

---

[1] James Douglas, *Notes on the Development of Phelps, Dodge & Co.'s Copper and Railroad Interests*, January 1906, 13. Reprinted by Frontera House Press, Bisbee, Arizona, 1995, with introduction by Gary Dillard.

[2] Roberta Watt (Troxell), *History of Morenci, Arizona*, Master's thesis, University of Arizona, 1956, 9.

[3] Willis F. Dunbar and George S. May, *Michigan: a History of the Wolverine State*, 1980, 479.

[4] *The Detroit Free Press*, obituary, January 3, 1875.

At Ward's death of apoplexy, which occurred on January 2, 1875 on a Detroit street as he walked to attend a session of the Superior Court where he was to appear as a witness, the president of the Detroit Board of Trade made this announcement:

> No man among us probably more thoroughly exemplified the grand characteristics of American institutions than Mr. Ward. I well recollect when his entire means consisted of an interest in the schooner *Harrison*, after which he brought out the steamer *Huron*, which, through his indefatigable energy and tireless industry, became the mother of a vast fleet of steamers which dotted the northern lakes from west to east; and later, when it became evident that railways would compete successfully for the carrying of passenger traffic, at the expense of steamers, his teeming brain brought out a scheme of iron works which has developed a city of 2,500 people and their vast industries in the iron manufacturing interests of Wyandotte (Michigan). Later, the North Chicago rolling mills, blast furnaces and steel works were established, and among other enterprises traceable directly to his master mind may be mentioned the Milwaukee rolling mills and furnace, the Frankfort rolling mills and furnace, and recently an enterprise for the manufacturing of plate glass, near St. Louis, with a capital of more than a half million dollars.[5]

It is interesting to note that among the many accomplishments of Captain Ward written about at his death, nothing was said about his involvement in the Detroit Copper Mining Company. It is probable that he never told those associates on the Board of Trade about the mines. It seems to have been in his character to not talk about his accomplishments. As an example, he was reported to be a benevolent man who seldom alluded to his works of charity even to his most intimate companions. We do not know if he followed any established religion, but he was a notable supporter of religion. A week before his death he had given $1,000 to a church on the west side of Detroit, a *considerable* amount of money in 1875.

Ward was well-loved by those who knew him. When word of his death was out, the flag at Detroit's city hall was flown at half mast. At Saginaw, a hundred miles north, the *Saginaw Courier* reported: "The buildings of the Flint & Pere Marquette Railway Company were tastefully draped in mourning yesterday in respect to the memory of the late President of that road. The engines and trains also bore similar somber-hued symbols of mourning and respect." A lawyer, upon being told of Ward's death, said: "There is gone the man from whose hand came the bread of hundreds, perhaps thousands." Another said that Captain Ward had "by great personal

---

[5] *The Detroit Free Press.*

sacrifice kept some of the extensive enterprises moving this winter that poor men might live."[6]

In 1875, William Church acquired options on all the Detroit Copper Mining Company stock from the Eber Ward estate and,[7] in 1880, with his brother, John Church, and brother-in-law Edward Whiting, left his home in Colorado for the site of the DCMC claims at Camp Joy, Arizona Territory.[8] That trip was made at one of the most perilous periods of the Indian Wars, and it was not the first time Church had tempted fate. Born in Quincy, Illinois and employed as a young man in Chicago, he left for Colorado in 1866 with a party of friends at the age of 21. At that time, the danger from Indian attacks was so great the military had required that 40 armed men accompany every party of civilians attempting the trip. According to Frank Hall in his *History of the State of Colorado*:

> Mr. Church went to Arizona, making this journey also by team, with but two companions, at a time when the hostiles of Geronimo's band were killing people both before and behind the three lone tourists traveling by wagon. They sometimes camped at ranches where the whites had just been murdered by the Indians, and the "close shaves" they had for their lives made this the most exciting experience of Mr. Church's life.[9]

Dr. James Douglas, who would later become president of Phelps, Dodge & Company, was another one of those remarkable men of which Phelps Dodge would have much more than its share. He had become acquainted with William Church as the result of Church's somewhat audacious approach to William E. Dodge, Jr. This meeting had occurred at Number 11 Cliff Street in New York City, the offices of Phelps, Dodge & Company. Its was the first visit that both Church and Douglas had made to that office. Douglas was there on another matter but would soon be drawn into the business for which Church was in attendance. Douglas, for whom the town of Douglas, Arizona would later be named, described his visit with William Dodge. The discussion had turned to the subject of Church's visit earlier in the day:

---

[6] *The Detroit Free Press.*

[7] Norman Carmichael and John Kiddie, *Transactions, American Institute of Mining and Metallurgical Engineers, Vol. LXX, 1924,* "Development of Mine Transportation in Clifton-Morenci District," 831.

[8] Roberta Watt (Troxell), 22. Also, *Transactions of the American Institute of Mining Engineers, Vol. XLIX (1914),* "Development of the Round Table at Great Falls," Arthur Crowfoot, 425-26. In this paper, Arthur Crowfoot, who was concentrator superintendent at Morenci, makes note of a concentrating table improvement made by one John Church at Tombstone in 1883. It is not known if this John Church was William Church's brother, but the timing of the event and the fact that there was no further mention of John Church with respect to the Morenci operation, makes it an intriguing possibility.

[9] Frank Hall, *History of Colorado,* The Blakely Printing Company, Chicago, 1895, 499.

> During our conversation Mr. Dodge said that a handsome, attractively frank man had that morning entered the office, and, without any introduction, asked for $30,000 with which to build copper works on a mine near the Longfellow.[10] The extraordinary impudence of the demand would have been resented but for the genuine sincerity and absence of all blushes or exaggeration on the part of the applicant. He was a Mr. William Church. He had secured undivided control of a number of claims near Clifton, adjacent to the Longfellow, by getting options on all the stock of the Detroit Copper Co. He wanted the money to develop his mines and erect smelting works. As Ansonia was treating Copper Queen copper, and the subject therefore interested Phelps, Dodge & Co., and Mr. Church had not been summarily dismissed — did I know anything about his mines?[11] I told Mr. Dodge what I knew from personal observation of the Queen, and what I knew from reliable information of the Longfellow. My story so impressed him that he asked me what I would charge for a report, on Church's property, as the information would be of value to them, even though they did not accede to Mr. Church's request.

In that way, two giants of what would be the new Phelps, Dodge & Company copper mining venture, a first-time experience for the company, were brought together. "Thus Church and I were strangely intermingled during our first introduction to Cliff Street," Douglas wrote, "and Phelps, Dodge & Co. took their first plunge into copper mining."[12]

But who were the men who ran that New York City firm that Church had considered important enough to travel 2,500 miles to see — and what kind of firm was it?

Anson Greene Phelps (1781-1853) was the founder. After being orphaned at the age of 10, Anson Phelps lived in a foster home and at age 14 was apprenticed to a saddler in Hartford, Connecticut. He later established his own shop where he sold objects made of leather and began the export-import mode of business. He sent saddles to Charleston, South Carolina and received back cotton which was sent to New York.

Phelps married Olivia Egleston (1784-1859) in 1806,[13] and with an eye for expanding his business, moved to New York and took in a partner, Elisha Peck. They called their firm "Phelps & Peck" and entered the arena of international trade. Peck established himself in Liverpool where he received cotton shipped to him by Phelps and in turn sent back metals which Phelps

[10] This also has been said to have been a request for $50,000. James Colquhoun, *The History of the Clifton-Morenci Mining District,* London: John Murray, 1924, 11. In any case, it was a considerable amount for the time.

[11] Ansonia, a plant in Connecticut, manufactured items from brass and copper. A sister firm, the Ansonia Clock Company, made clocks that are now prized by collectors of antiques. The town of Ansonia, Connecticut carries on the name of Anson Greene Phelps.

[12] James Douglas, 12. Frank Hall, 499.

[13] Phyllis B. Dodge, *Tales of the Phelps-Dodge Family,* New York Historical Society, 1987, 6-9.

could sell or make into useful objects at the Ansonia, Connecticut metal works that he had founded.

The Phelps and Peck partnership was dissolved following the disastrous collapse of their cotton warehouse on May 4, 1832 in which seven men had died. Included among those who perished was the clerk who was to marry Caroline, then the youngest of the Phelps daughters.

In 1834, Phelps took his two sons-in-law, William Earl Dodge (1805-1883) and Daniel James (1801-1876), into the company. Daniel James had married Elizabeth, the eldest daughter, and William Dodge had married Melissa who was two years younger than Elizabeth. The firm was called Phelps, James & Co. in Liverpool and Phelps, Dodge & Co. in New York.[14] It was to this Phelps, Dodge & Co. office at Number 11 Cliff Street, New York City, that William Church and James Douglas coincidentally appeared on that same memorable day in 1881 and where they separately held discussions with William Earl Dodge, Jr. (1832-1903) that would have ramifications far beyond their possible imaginings.

A trip to New York City and back in those days of the Wild West, such as William Church and James Douglas had made, had to have been somewhat of a story in itself. Douglas continued:

> I met Church by appointment in Lordsburg. We drove across the desert and divide into the valley of the Gila at a post office town, which has since completely disappeared, called Richmond; thence into Clifton over a stage road, which followed approximately the present railroad route.

At the time of their survey of Church's claims, which were adjacent to the Longfellow Copper Company workings, the Lesinskys had been working the Longfellow Mine for about eight years. Douglas reported that "Every courtesy and facility was extended (by the Lesinskys) to Church and myself."

Following his inspection of the site, Douglas reported favorably on the Morenci ore body and also on a Bisbee, Arizona property with the result that Phelps, Dodge & Company, in a decision made final on May 21, 1881, acquired an interest in the Detroit Copper Mining Company and William Church now had the capital to begin construction of a copper works. In the next month, Phelps Dodge & Company, the New York City mercantile firm, would also acquire the Atlanta claim in Bisbee.

As he considered how he would build his copper works, Church had three choices: (1) he could build his smelter at the mine and pump water to it

---

[14] Phyllis B. Dodge, 42-44.

from the San Francisco River which was seven miles distant and 1500 feet lower, (2) he could construct the smelter at the river and transport the ore by rail, or (3) he could transport the ore by wagon (with the smelter at the river). The cost of the first two options forced Church to choose the last. About three miles down-river from Clifton, he erected two 36-inch furnaces, a store, office and dwellings. The furnaces were run by water power.

Church expected to be producing copper by the autumn of 1881, but the summer monsoon rains delayed construction. It wasn't until the spring of 1882 that the first furnace was blown in, but due to metallurgical problems very little copper was made. At this juncture, one of James Douglas' assistants in Pennsylvania, Young Zogbaim, was sent to help Church with the metallurgical problems but was drowned in the San Francisco River.

In the next spring, 1883, James Douglas sent a Mr. Hyams to Clifton to "analyze the charge and make up proper furnace mixtures."

The metallurgical problems were not the only ones Church had to contend with. During that time, there was, as a practical matter, no rule of law in Clifton and Morenci. Church was held up three times by masked gunmen. Geronimo was another problem, and he was a big problem. Geronimo's warriors weren't only attacking the miners. This was an all-out war against the intrusion of the European-Americans. They also attacked ranches in the York Valley and in the Eagle Creek area. In an unpublished manuscript, Frederick Remington, the great painter and sculptor of Western subjects traveled through our area in the 1880s and recorded his observations. In one piece, he wrote about the renegade Apaches who had been raiding, "killing horses and cattle and terrifying the settlers along Eagle Creek."[15] A band of 500 Apaches attacked the smelter works, raided the mines, drove off the company's and contractor's livestock, and killed eleven men.[16] All operations were then stopped for several months.

Church now reconsidered his options. He could build another smelter at the mines, but that would entail constructing a water pipeline. Or he could build a railroad which would be safer from Indian attacks since they were attracted to the livestock. On a comparison of costs, Church chose the pipeline and the smelter was constructed up at the mines. Church's careful calculations of the costs of his options may have partly been the reason that prompted Douglas to write: "He (Church) was timid in making expenditures and was inclined to favor cheap construction. But he was a wonderfully

---

[15] Harold McCracken, editor, *Frederick Remington's Own West,* "How an Apache War Was Won," previously unpublished manuscript by Frederick Remington, The Dial Press, New York, 1960, 61.
[16] Frank Hall, 499.

economical and faithful manager — as true to his workmen as to his fellow shareholders."[17]

James Colquhoun, a young contemporary of Church, told the story of the geologist who had visited the Detroit Copper Mining Company mines and had impressed one of the DCMC McLeans (Alec McLean) with the names of the various rocks to be seen in the mine. When McLean tried to so educate his boss, Church, he was told he should know there were only two kinds of rock, smelting ore and waste rock, and McLean "was not to forget it." According to Colquhoun, Church had even less regard for geologists than for sulfide ore, "which meant little indeed." But this final comment revealed how much he admired Church:[18]

> Of fine presence, highly educated and energetic to an uncommon degree, he would have been an imposing figure in any assembly. He was a big man and came very near to being a very great man....He had some of the faults of a headstrong nature, but his virtues made his faults look small, and in his day he did more than a man's work and was, as he deserved to be, universally respected.

After 1887, perhaps missing the comforts of city life, Church lived in Denver and only made visits to the mines which were being managed by a Mr. Sweeney. Until he sold out to Phelps, Dodge in 1897, he continued to serve as president and general manager of the Detroit Copper Mining Company. After purchasing Church's interests, Phelps Dodge embarked on a modernization program and, until 1917, continued to operate as the Detroit Copper Mining Company. Construction of the West Yankie concentrator was one of its first building projects. This was needed to treat the sulfide ores that were then being encountered.

Church received $1,800,000 for his mine holdings. He built a beautiful home in Denver that was known as "the Castle," engaged in other business arrangements and gave to philanthropic and educational enterprises. A firm believer in kindergartens before these became a part of the public school system, he established four kindergarten schools in Denver, paid for the training of the teachers and was the founder of the Denver Kindergarten Normal School.[19]

William Church died on May 25, 1901 of apoplexy in his Denver home known as the castle. During World War II, the Red Cross used the beautiful house as lodgings for servicemen. In August 1965, the castle fell victim to the wrecking ball[20] to make way, we were told, for apartments.

---

[17] James Douglas, 14-15.
[18] Colquhoun, 45-46, 71.
[19] *Denver Republican,* May 26, 1901, 8.
[20] *Rocky Mountain News,* August 6, 1965.

Anson Greene Phelps (1781-1853) William Earl Dodge (1805-1883) Daniel James (1801-1876)

Anson Greene Phelps began the company that would become the Phelps Dodge Corporation. His two sons-in-law, William E. Dodge and Daniel James, were business partners. *Courtesy Phelps Dodge Corporation.*

On May 4, 1832, the Phelps and Peck cotton warehouse collapsed, leading to the dissolution of the partnership and the formation of a new one called Phelps Dodge & Company. *Courtesy Phelps Dodge Corporation.*

William Church (1845-1901), president of the Detroit Copper Mining Company. With a loan from William E. Dodge, which he used to build a smelter near Clifton, William Church brought Phelps, Dodge & Company into copper mining.
Courtesy Colorado Historical Society.

In 1882, William Church of the Detroit Copper Mining Company built this smelter on the San Francisco River at a point about three miles downstream from Clifton. The operation would be fraught with difficulty. *Courtesy Phelps Dodge Corporation.*

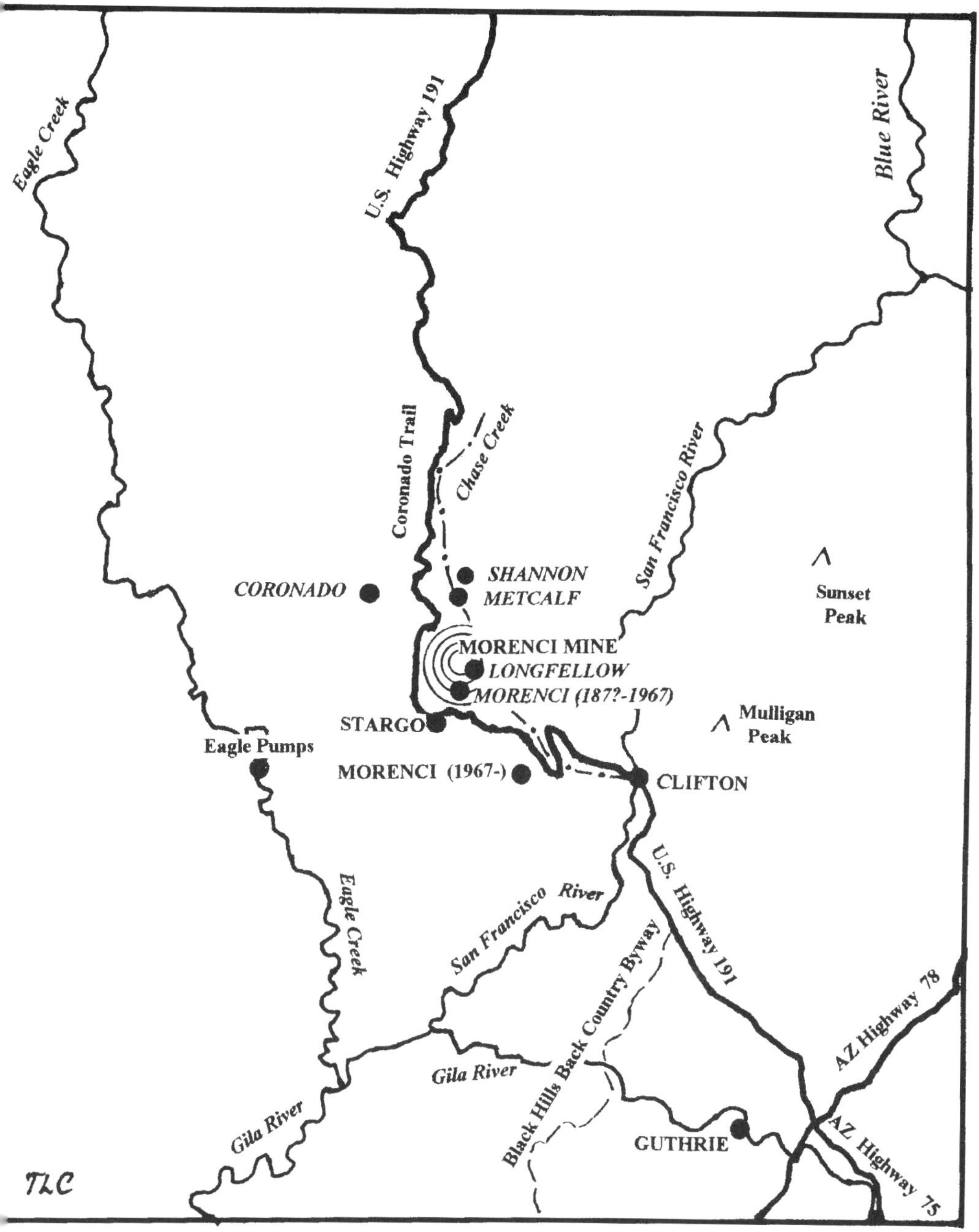

The communities of Metcalf, Shannon, Coronado, Longfellow and original Morenci disappeared long ago. Morenci was the last to go, being dismantled from 1965 to 1982 and rebuilt about three miles south.

Dr James Douglas who recommended to the Phelps Dodge partners that they enter into copper mining at Morenci and at Bisbee. *Courtesy Phelps Dodge Corporation.*

Old Morenci with the Detroit Copper Mining Company smelter in the right foreground, 1897. *Courtesy Roberta Troxell.*

The Detroit Copper Mining Company smelter in Morenci, Arizona Territory, about 1897.
From the Merna Robison collection.

This man, identified as Dick Stevens, delivers bread to residents of Old Morenci.
From the Merna Robison collection.

8

1883: James Colquhoun, the right man at the right time....TLC

In his biographical sketch written in 1950 for a Phelps Dodge Morenci Branch quarterly magazine, *The Prospector,* James Colquhoun wrote: "The chance to make a fortune, it is said, comes to every man once in a lifetime, but he must be ready and prepared to grasp it with firm hands when it does come."[1]

Born to a poor family in 1857 in the outskirts of the village of Chapelhall, Scotland, Colquhoun had followed that advice. He had been put to work at an early age and gradually rose to the position of bookkeeper for the Monkland Iron and Steel Company, one of the oldest iron firms in Scotland.

Knowing that he had to educate himself well if he wanted to rise in the company, in a nearby town Colquhoun attended night classes in a school under the control of the government where he focused on the study of chemistry and metallurgy with lesser emphasis on geology, magnetism, electricity and mining. His father, a highly educated man, advised that he also study the works of the historian Macaulay and other authors as an aid to composition.

At the Monkland Iron and Steel Company, he compared what he had learned in theoretical metallurgy with the practice of metallurgy. Away from the more serious work, he read the American authors Mark Twain, Artemus Ward and others, acquainting himself with American humor and with America itself. As a leading participant in a debating society, he gave his friends the opportunity to enjoy choice passages from the American humorists.

In 1883, Colquhoun was offered the position of bookkeeper with the Arizona Copper Company. From its headquarters in Edinburgh, Scotland, the company had recently purchased mines in far off Morenci, Arizona

---

[1] James Colquhoun, biographical sketch written for *The Prospector,* 1950. In this he listed the year of his birth as 1858, which was probably an error in typing. His son, James Clifton Colquhoun, lists the birth date as October 29, 1857, a date which grandson James Colquhoun believes is correct.

Territory. Colquhoun would be going to that Arizona mining operation. The mines were the Longfellow, and the Metcalf and Coronado groups. Their company town would be located north of Morenci and would be called "Metcalf," named after Robert Metcalfe who had been the discoverer of several Morenci-area mines.

Colquhoun sailed from Glasgow to New York aboard the *Ethiopia.* In August 1883, he rode the narrow gauge (three-foot) railway from Lordsburg to the rails' end at Guthrie. From there, he rode the stage coach some seven to eight miles in a north-northwesterly direction, much of it over rough country. This brought him to Clifton where he learned the Apaches were on the warpath and the town was dominated by the desperado Kid Louis and his band of outlaws.

Based largely on the legendary gunfight in the OK corral, Tombstone has become a haven for tourists who want to be in the place where the West was truly wild; but it would be easy to argue that Clifton, Morenci and Metcalf were even more appropriate examples of those desperate days. It was a time when residents of the three communities considered rustling, holdups, shootings and attacks by the Apaches, inspired by their leader Geronimo, as something to be expected.

Into that scene arrived James Colquhoun, a 25-year-old Scotsman with a fond appreciation of American humor. During the first year of his employment he lived in a primitive hotel in Clifton. He was hired by the Arizona Copper Company as a bookkeeper, was put to work as an assayer and soon experienced a dramatic career change — the sort of thing that probably would have touched his sense of humor. Before much time had passed, instead of working as a bookkeeper or an assayer, he was assigned a monumental responsibility. The task: nothing less than to save Arizona Copper Company from an impending complete economic collapse.

Lawlessness in Clifton and Morenci was rampant during that time. Colqhoun, a gifted writer, wrote of those times in two books about Clifton-Morenci. In his *The History Of The Clifton-Morenci Mining District* Colquhoun relates that, over a three month period, an average of one man a day was murdered in Clifton. Kid Louis and his gang ruled the town, rustling livestock, robbing and killing. Neither he nor any of his gang members were ever convicted of any crimes. If one was arrested and brought to trial, Kid Louis and the rest of the gang, with weapons of their trade strapped on, would attend the trial to make sure "justice" was done. The jury's verdict was always "not guilty."

Ironically, Kid Louis was shot out of his saddle one day in what at the critical moment was probably a case of mistaken identity. He had been riding in Ward's Canyon when he encountered a posse composed of Mexicans who were old soldiers and good shots. The posse had been organized by a Mexican rancher and was in pursuit of cattle rustlers when it unexpectedly came upon Kid Louis. Seeing his way blocked and suspecting, no doubt, that they were after him, the Kid attempted to ride through the posse with guns blazing. The posse's aim, it turned out, was better than his and the reign of terror ended.[2]

Arizona Copper was operating under extremely difficult financial conditions when Colquhoun, in 1889, was given the job of president. Colquhoun described it this way: "The history of the district up to 1892 may be described as one continuous struggle for existence."[3]

In 1893, with the company facing certain financial ruin, he managed to convince the mortgage company that they would be better served to defer payment on their loan with the hope that they might receive something from a leaching process, that he had invented and wanted to implement, rather than to simply foreclose with the certainty of no return at all.

To embark upon this new copper-producing process required some courage. Colquhoun understood that probably better than anyone else. "No one," he wrote, "could approach such a task as the design and construction of a plant suited to a new process of leaching copper ore without knowing that he was treading on dangerous ground."[4] But of course, it worked, saving the company from bankruptcy.

Then, in 1896, with the higher-grade ores mined out, Colquhoun had a concentrator constructed that was designed to handle the low-grade sulfide ores. He thus became the first to produce copper successfully from low-grade porphyry ores.[5]

---

[2] James Colquhoun, 26-30.

[3] James Clifton Colquhoun, son of James Colquhoun, in his *James Colquhoun 1857-1954 The Man and His Work, p.11,* writes: "he was asked (in 1889) to take charge of the mines as well as the smelter. By the end of 1892, when the ACC was on the verge of closing down, he was asked to take over the general management of the Company." James Colquhoun himself, in a letter written January 12, 1950 on Empress Hotel, Victoria, B.C. stationery decorated with a crown as logo, wrote: "Colquhoun started work as a bookkeeper in August 1883. About 1888 he became president of the Arizona Copper Company and Arizona and New Mexican Railway. In 1907 he resigned his position on account of poor health.".

[4] James Colquhoun, *The History of the Clifton-Morenci Mining District,* John Murray, Albemarle Street, W. London, 1924, 49.

[5] James Clifton Colquhoun, p. 29. He points out that an authoritative work had credited Utah, in 1905, with the first production from a porphyry mine and then gives ample proof to show that it was the Arizona Copper Company that first produced copper from the porphyry ores and that it had done so with a pilot plant in 1896 and with a full-size porphyry concentrating plant, the No. 3 Concentrator, in 1898.

We now jump forward nearly one hundred years to October 12, 1993. On that day we were favored with a memorable visit. James Colquhoun, grandson of the James Colquhoun of Clifton-Morenci, and his wife, Sarah, toured the Phelps Dodge Morenci, Inc. operations. They provided us with engaging conversation and answered our many questions about the Colquhoun family, which we then learned was pronounced "Koh-hoon." It was a most enjoyable day.

In our conversation with his grandson, we discussed James Colquhoun's World War I trip to Russia where he went to run a copper mine in the Caucasus, an undertaking that he later described in his book, *Adventures in Red Russia.*

When the Bolsheviks took control of Russia in 1917, they began to interpose themselves into the operation of the mine. Colquhoun could not ship copper, which meant there would be no money to pay the workers, and there was a shortage of food.

It was obvious that he then had to cease operation — there was no other option. But the new rulers and the workers opposed a shutdown. Finally, in open defiance of the authorities, he announced the shutdown which took effect on December 31, 1917 and made preparations to send his staff out of the country by way of Vladivostok. They left but were captured by Tartars and were turned over to the Turks as POWs where they would remain until the Armistice.

Having business matters to clear up, James Colquhoun and his mine superintendent had stayed behind. Later, they began to make their way to the north. It was a difficult journey, fraught with hardship. Finally, they reached the port of Murmansk from which they left the country in June 1918. After many months of absence, James Colquhoun one day walked back into the office in Scotland, looking altogether as if nothing of consequence had happened.

Colquhoun absorbed well the results of his father's advice — to study the works of Macaulay and other authors as an aid to composition. He wrote six books:

Captain W.W. Colquhoun
The History of the Clifton-Morenci Mining District
The Early History of the Clifton-Morenci District
Adventures in Red Russia
Birth of the Porphyry Coppers
Chamberlain Averts Armageddon

In later years, Colquhoun moved to Victoria, British Columbia and, in spite of having had to retire from the Arizona Copper Company "for poor health" way back in 1907, was fit on into his 90s. He died at the age of 96 on June 17, 1954. In the biography written by his son, we learn that Colquhoun had contracted typhoid fever in 1887 shortly after his return to Clifton from a trip to Scotland where he had married Elizabeth Scott Wallace. This was " a long illness which took its toll; from then on he felt it necessary to escape from the Clifton climate for part of each year."

The illness could have been the reason he felt it necessary to retire for poor health in 1907; but knowing that he would later be energetic enough to take on the Bolsheviks in the Soviet Union, one suspects that, just perhaps, by 1907 life in Clifton-Morenci had gotten to be a bit too tame for James Colquhoun. In Wards Canyon, Kid Louis, with a revolver in each hand, had tried to run through a posse organized by a Mexican rancher and was killed; Geronimo had surrendered in 1886; and the company, through his efforts, not only had been saved from ruin but, on August 6, 1895, began paying dividends and did so thereafter on a regular basis. Now that the West was essentially tamed and Arizona Copper was thriving, James Colquhoun faced a remainder of a career in routine management. That was probably not his cup of tea.

It would be fair to say that James Colquhoun was much loved by the people of Clifton-Morenci. Each year after his retirement, he sent a $100 gift at Christmas time to be used for gifts for Clifton's children. On one occasion, following their return from a summer's trip to Scotland in about the year 1900, they were given a grand reception. The townspeople were at the depot to meet them and Clifton's homes and businesses were decorated with flags, streamers and Chinese lanterns. Firecrackers were set off in the evening.

The Colquhoun house was located approximately where Highway 191 now runs past the parking lot at the Clifton train station. In 1913, about six years after the Colquhoun's moved from Clifton to Britain, it was purchased by Clifton's east side town baker, Frank Salerni, and his wife, Jenny, and moved across the San Francisco River.

In April 1993, the Salerni's great grandson, Ralph Guida, and his wife, Vicki, came to Clifton from Orange, California to view the home. James Colquhoun, the grandson of the savior of Arizona Copper, and his wife, Sarah, also viewed the home during their tour.

James Colquhoun of Arizona Copper and Elizabeth Colquhoun had two sons — William Wallace (Willie) and James Clifton.

On September 25, 1915, during World War I, Willie Colquhoun, a Clifton boy, was killed in the trenches at the Battle of Loos in Belgium. Also during that war, from 1914 to 1918, Willie's brother, James Clifton, who had been born in Clifton on December 1, 1893, served as a captain in the Highland Light Infantry.

From 1922 to 1931, James Clifton Colquhoun was director of The Cornish Tin Smelting Company, Cornwall. Profiting from his father's discovery of a way of reducing tin in the slag from 4% to 1%, he was "mistakenly credited," he wrote, "with great success as a tin smelter."

James Clifton Colquhoun later was made chairman of Manganese Bronze and Brass, a position he held to 1960. At first the factory was located in London, making ships' propellers, including those for the luxury liner RMS Queen Mary that is now a floating hotel in Long Beach, California, and oilite bearings and other things. A bomb fell on the factory during World War II, and the operation was moved to Liverpool.[6]

In 1970, James Clifton Colquhoun wrote a biography of his father. In that work, he explained that the elder James Colquhoun had, in 1896, been the true developer of the concentrating process which made possible the mining of porphyry ores, a fact that had been mistakenly credited to others in the mining industry. "No other porphyry mine," he wrote, "was developed in this way on any scale until 1905...." James Clifton Colquhoun died February 9, 1977 at the age of 83.

James Colquhoun, the grandson, our visitor of October 12, 1993, retired in about 1991 from the position of agricultural adviser. He and Sarah have a home in Wales on the Atlantic shore and are enjoying a bit of travel in their retirement. It is our fond hope these travels will bring them back for more visits to Arizona.

---

[6] From a conversation with James Colquhoun, the grandson, October 12, 1993 and his letter to Ted Cogut dated December 2, 1993

James Colquhoun (1857-1954) "The Great Innovator." In 1893, he saved the Arizona Copper Company from financial ruin with his leaching process. In 1896, he became the first to produce copper on a production basis from low-grade porphyry ores. *Courtesy James and Sarah Colquhoun.*

A scene such as would have appeared to James Colquhoun in 1883 when he got off the train at the track's end at Guthrie. *Courtesy Roberta Troxell.*

On October 12, 1993, James Colquhoun and his wife, Sarah, toured the Morenci Mine. James Colquhoun is the grandson of President James Colquhoun of the historic Scotland-based Arizona Copper Company of Clifton-Morenci. Pictured in the heavy-duty truck shop are (left to right) Sarah Colquhoun, Craig Sircy, James Colquhoun, Ted Cogut. *Morenci Copper Review, November 1993.*

Clifton held a grand welcoming for Mr. and Mrs. Colquhoun and sons James Clifton (standing in front of his father) and William Wallace (Willie) upon their return from a vacation trip to Scotland. The townspeople decorated the Colquhoun house with flags and a "Welcome Home" arch. *Courtesy Greenlee County Historical Society.*

The Clifton rail yard in the early 1900s showing the Arizona Copper Company staff houses. The house at the far end of the row is the Colquhoun house. Note the absence of the Clifton railroad station which was not to be built until 1913. The three-track rail system would accommodate trains of both the baby gauge with 20 inches between rails and the narrow gauge of 36 inches between rails. *From the Merna Robison collection.*

9

The Historical Society receives a unique gift from PDMI..........TLC

On one otherwise routine day in 1990, Fred Menzer, Phelps Dodge Morenci Inc.'s chief geologist, walked into my environmental services office and asked if I might be interested in acquiring for the Greenlee County Historical Society a plaster-of-Paris cast, a relief model, of the Morenci Mine. Without hesitation I said "yes." Fred's question came while I was serving as the Historical Society's president, a job that the Society had assigned to me the previous year in a most uncharacteristic moment of weakness.

The relief model would obviously be an appropriate acquisition for the Society's museum. With the great assistance of Harvel Cosper, I had managed to acquire historic horse-drawn agricultural implements for the museum. Agriculture had long been important in the county, but the area's great copper ore body was world renown, and the copper produced from it was in fact at the core of the county's economy. I had long been looking for any additional artifacts related to copper mining.

I hustled over to the geology department to look at the casting. And what a piece of work it was! On it, the Clifton-Morenci mountains and canyons were depicted in relief and in various colors. Interestingly, the great Morenci open pit was not shown. The relief model obviously was made prior to 1937, the year that mining by open pit methods began in Morenci.

I saw that the casting had been made by the Detroit Copper Mining Company. This meant that it had to predate the year 1917 for that was when Detroit Copper's name was changed to Morenci Branch, Phelps Dodge Corporation — three-quarters of a century in the past! Without doubt, I was looking at an important piece of local history.

Years before, the casting had been stored away in the department's storage rooms where it stayed, out of sight and forgotten. It had first been stored in Old Morenci and had been moved to the new geology buildings when that move was made in about 1967 in conjunction with the relocation of the townsite to make room for open pit mining.

Mounted on a heavy oak base with slightly tapered legs that were so thick they were more than a foot wide at the top, the plaster-of-Paris casting was encased in a raised glass cover. At its thickest, where the mountains were being depicted, the plaster-of-Paris was about a foot and a half deep. The terrain was colored in various hues to identify geologic formations. The railroads of the time had been carefully drawn on the casting in black as were the buildings of Metcalf, Old Morenci and Clifton. Downstream along the San Francisco were the smelters of the Shannon Copper Company and the Arizona Copper Company. These were on opposite banks of the river and south of Clifton's populated area — the Arizona being 2.2 miles down river and on the east bank, at the site that was called "Smelter Hill," and the Shannon on the west side and a little closer to town. As you looked for Old Morenci and Metcalf up Chase Creek Canyon, you saw that the inclines from the mines were also shown.

I approached Jim Madson, then PDMI manager, to ask permission to acquire the casting for the museum. My first thought was to begin the conversation with the remark that, unlikely as it might seem, I thought Ohio State could beat Michigan in football — Jim was an Ohio State alumnus — but not being football season, that slight attempt at humor might seem inappropriate. Instead, I brought up the mine's environmental status (my department), which I casually noted was great, and then, in an equally casual way I asked for the casting. Like all the Morenci managers that I had known or heard of, Jim was a solid supporter of the community. Jim quickly gave his consent to that gift and also of some copper mineral specimens.

Phelps Dodge had provided other gifts for the Society's museum, chief among them was the historic Detroit Copper Mining Company safe which was so heavy it could not be brought into the museum proper for fear that it would plunge through the wooden floor. It was placed, instead, on the museum's concrete back porch.[1]

From the fact that the Arizona Copper Company smelter on Smelter Hill was shown on the casting, one could narrow down the likely date of its completion. Erection of the smelter began in February 1912 with construction of all elements completed in February 1914. Construction was far enough along to allow for the first copper pour to take place in October 1913. Of further interest was the fact that the Detroit Copper Company name

---

[1] The safe had previously been proudly displayed by John Bolles in his manager's office. With his transfer to the Ajo operations, Carl Forstrom became manager. Carl donated the safe to the historical society.

became history in 1917. At that time, a reorganization changed the name of the operation to "Morenci Branch, Phelps Dodge Corporation." This provided us a five-year window within which the relief model had to have been made. But also, depicted on the model with black paint at the smelter's location was a certain amount of slag that one could estimate would take a couple of years of smelting to produce. So the relief model was made, by that calculation, during the year 1915. That was when all mining had been underground and more than 20 years before work had begun in digging out the great open pit of the Morenci Mine.

Moving the casting to the museum, I quickly saw, would not be one of those small chores that one might accomplish in a few moments. First of all, it was fairly large — about six feet square — and it was heavy. I asked the PDMI research department to calculate the weight of the plaster-of-Paris that made up this fantastic work of art. This estimate would depend on my guess as to the thickness of the casting which could not be precise because of the mountains and valleys it depicted. But there was no doubt that it would be heavy. Their estimate was that it would weigh about 1500 pounds. And of course there was also the weight of the thick oak base.

Given its size and weight, could it even be moved without destroying it? In that previous move that must have been made in the 1960s, small pieces had been chipped away from the corners — but the amount of damage had been inconsequential — I feared that whole sections might break away in another move. I asked Jim Ballard, head of the carpenters/painters department, to meet me at the geology department and give me his appraisal of the feasibility of a move. The carpenters and painters are another of those groups at Morenci, like the maintenance crews, the electricians, the drilling and blasting crews, the surveyors, the draftsmen, to name only a few, who are not usually recognized but are, in my opinion, of more basic importance to the operation than many of those who frequently bask in the limelight. The carpenters and painters and those other crews keep the operation running smoothly and make possible the mining production records.

Jim Ballard said the casting could be moved. "But how?" I asked. Would it have to be cut in two? It was a terrible thought but one that I had been mulling about, thinking that to cut the casting in two was the only way it could be done. I was thinking that in addition to the ordinary mechanics of the move, there were the complications of carrying the casting up the steps leading to the museum from Chase Creek Street and then the turning of the object on edge as it was brought through the museum doorway. That would not be easily done. If it were cut in two, afterwards, presumably, the two sections could be rejoined and Dan Smiley of the drafting department would be nice enough to retouch with paint the chipped edges as he had already done in repairing the minor damage that had occurred in the previous move some 22 years earlier. But rejoining it could not be done without some damage that I feared couldn't be perfectly repaired.

It would not be done that way, Jim said. Instead, he would remove the casting from its oak base, make a crate, put the casting in it and move it into the museum in one piece, and probably with no damage at all. And that was what they did. After moving the oak base to its assigned place in the museum, the carpenter/painter crew gathered around the crate that held the casting, lifted it out of their truck, carried it edgewise up the museum steps, and placed it upon its oak base. And they did it with no damage whatsoever.[2]

There was only one sort of unexpected outcome. Remembering the concern over the weight of the Detroit Copper Mining Company safe, I had asked Jim Ballard to bring two six-foot-long two-by-six planks to place under the model's legs so as to bridge the several floor joists and spread the weight over them. After the model had been placed on the planks, the weight bowed the planks to very near their breaking points. But that was no cause for alarm, for Jim then brought two three-by-twelve planks to replace the two-by-sixes. These perfectly carried the weight.

I went back to the museum one afternoon on a day when it was not open to the public. Alone in the quiet amongst Greenlee County's historical artifacts I sought a mental picture of that time — the time when the casting was made. That, we had decided, would have been 1915. Unless it had been during the period of the strike of that year, Morenci's miners would have been producing copper that was destined for use in the Great War; and in two more years the United States would be in the war and a number of Morenci's and Clifton's finest would have been far away, serving on the Western Front, an event and place from which, perhaps, some would never return.

That plaster-of-Paris casting provided a perfect picture of what the terrain had been like before those spectacular 50-foot mine benches had been developed. High above Chase Creek Canyon, the miners would have brought the copper ore to the top of the inclines which were so perfectly marked on the plaster-of-Paris. Most of all, I marveled at the technology of the time, so well exemplified by those inclines down which the ore was sent to waiting baby-gauge locomotives. Several important advances in technology had been perfected at those mines; but for many, those fabulous inclines characterized the district. There were more of them — nine in all — than at any other mining district and among them were the world's longest and the steepest.

[2] In addition to Jim Ballard, six or eight of these gentlemen from the carpenter-paint shop carried the model into the museum: Mike Munoz, Eloy Cervantez, John Marsh, Frank Bailon, Billy Gonzales, Chuck Richard, Clint Norton, Bruce Stark, Charlie Carr, Royce Stephens, Jesse Smith, Charles Boling and Doug Day.

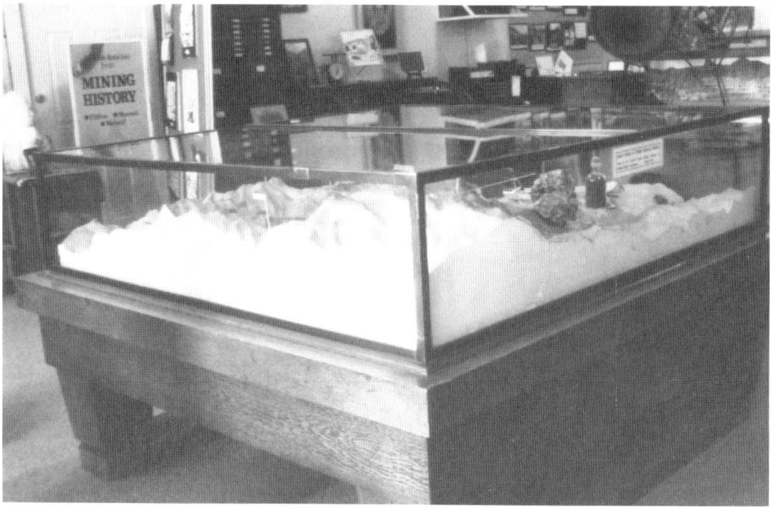

This relief model of the Clifton-Morenci District shows the mining area as it was a little more than 20 years before excavation for the open-pit mine began. It was made by the Detroit Copper Mining Company in about 1915. Donated to the Greenlee County Historical Society in 1990, it can be seen in the Society's museum on Chase Creek Street in Clifton.

Henry Lesinsky built this home in Clifton. Reinforced in its construction with slag, it withstood the powerful 1983 San Francisco River flood. Water rose well up into the house and destroyed a number of irreplaceable artifacts of the Greenlee County Historical Society which was then using the building as its museum. Believed to have been built in 1872, it is the oldest standing house in Clifton. *Photos by Ted Cogut, 1997.*

In 1990, at their new museum in the Spezia building in Clifton, Al Fernandez, Harriet Sweeting and Andres Padilla, distinguished members of the Greenlee County Historical Society, and Ted Cogut discuss items donated by Phelps Dodge that were used in the earliest days of mining. The Detroit Copper Mining Company safe behind Andres Padilla was too heavy to be placed on the wood floor inside the museum. In the 1970s-80s, Phelps Dodge Morenci Manager John Bolles proudly displayed the safe in his office.

Courtesy Phelps Dodge Corporation.

10

Spectacular inclines were used to bring ore down from the mountains...WCC

Most of the old underground mines in the district were located in extremely rough, mountainous terrain. Getting the ore from mine to market was an almost insurmountable task. The first step was to transport the ore from the isolated mine to the railroad or, before that, to the wagon trail.

As we look at the mining history of Morenci we see many "firsts" in the employment of new technology. One of these, dating way back to those earliest mining days, surely was the district's extensive use of inclines. Altogether, there were nine inclines — no other mining district that we know of had that many.

With the old-time mines of Morenci being located at or near the tops of mountains, that first step, getting the ore down the mountainside, was a big one. One tool used by many of the early miners was the "incline," which consisted of a steep set of mine-car rails dropping from the mine portal to the canyon below. A cable reel with heavy cable was located at the top of the incline. Mine cars full of ore would be hooked onto the cable and lowered down the incline to ore storage bins. The ore could then be dumped into railroad cars for transportation to the mills and smelters.

The Longfellow incline

The first incline in the district was the incline at the Longfellow Mine which was the main ore producer for the Longfellow Copper Company.[1] That incline was built in the 1870s and extended from the mine to the bottom of Chase Creek Canyon, originally connecting with a wagon trail and later, by 1879, to the Coronado Railroad. The first stage of the Coronado Railroad to

[1] The Longfellow Mine, which became well-known as a producer of high grade ore, was sold to the Arizona Copper Company in 1882 and finally abandoned around 1903. Its high grade ore was then about gone. Also, there had been a disastrous cave-in.

be built, which carried the ore from the bottom of the Longfellow Mine incline to the smelter in Clifton, was four miles long.

Up at the mountain-top mine, one-ton ore cars were rolled on tracks to the edge of the cliff, attached to cables, and lowered down the incline to the ore bins below.[2] The empty cars would then be hauled back up the incline. Part of the Longfellow incline passed through a tunnel about half way up the mountain. This was the only mine incline in the Clifton-Morenci District, and quite possibly in the world, that ran part of the way through a tunnel.

The Coronado incline

This incline was said to be the longest in the country. Built to handle ore from the Coronado Mine, this incline rose almost 1200 feet in elevation. The Coronado Mine was located in the rugged mountains above Coronado and Santa Rosa canyons. It was west of and high above Chase Creek Canyon. In the canyon directly below the mine lay the town of Metcalf. Original claims on the Coronado vein were located in 1874 by the Coronado Mining Company and later became the property of the Longfellow Copper Company.

At the mountain top, the incline connected with a little railroad, about a mile long, which ran from the Matilda shaft of the Coronado Mine at the head of Coronado Canyon. Baby gauge trains hauled ore from the Matilda shaft to the top of the incline. There, the ore cars were connected to a cable and lowered to the bottom, where connection was made with the Coronado Railroad.

People often rode the incline in those little cable cars. One day in August of 1913, The *Copper Era* screamed this headline: "NINE MEN KILLED INSTANTLY WHEN ORE CARS DASH DOWN CORONADO INCLINE." The headline continued: "Six Men Have Narrow Escape By Jumping At Critical Moment." The accident happened on August 13, 1913. The 15 men were loaded into two ore cars preparing to be lowered down the incline. As soon as the last car dropped over the knuckle at the top of the incline, the drawhead on the last car, which was attached to the cable, broke, sending both cars plunging down the incline. The *Copper Era* provided these further details of the disaster:

> The ground was plowed up, in some instances on both sides of the track, as the cars swerved from side to side in their mad race. Trees had been chopped off like kindling wood that stood several feet from the track, and all along were strewn

[2] Roberta Watt Troxell, *History of Morenci, Arizona,* Master's thesis, University of Arizona, 1956, 85.

bits of clothing, chunks of human flesh, pieces of skull and bone....The scenes presented at the morgue were ghastly. No words can describe the awful and multiple injuries inflicted upon the bodies by the cruel stones upon which they struck in the wild descent of the cars from the knuckle at the incline's top, to the place where the rails were left, and the cars and men together rolled for hundreds of feet down the steep mountain.[3]

The incline was used for 10 more years, until 1923, when the Coronado Mine closed for the last time. Rail from the incline was salvaged in the late 1930s, and three little locomotives were left isolated and inaccessible on the mountain near the Matilda shaft. They had all been there for about 100 years when Phelps Dodge finally brought them down on a drill road in 1990.[4]

The Shannon incline

A spectacularly steep incline located north and east of the town of Metcalf rose out of the east side of Chase Creek Canyon and up to the mines of the Shannon Copper Company. The mines were shallow underground workings located on a mountain overlooking Arizona Copper's Metcalf mines and the town of Metcalf. It was the steepest incline in the district. Parts of the old incline are still visible as this is written (1997).

Little mine cars filled with ore were drawn from the Shannon mines by mule train to the top of the incline. Each car was then hooked to cables and lowered down to storage bins. Narrow gauge trains of the Shannon Railroad pulled in under the bins. When the train was loaded, it proceeded down Chase Creek Canyon through Metcalf to the Shannon mill and smelter located on Shannon Hill in Clifton. The incline was also used to bring supplies up to the mines and the town of Shannon, located on the mountain near the mines.

Shannon sold the mines to the Arizona Copper Company in 1919 following an economic depression in the aftermath of World War I. All the mines around Metcalf, including Shannon, closed that year, never to operate again.[5]

---

[3] *The Copper Era,* August 16, 1913.

[4] For a complete description of the difficult task of bringing the locomotives down the mountain, see Chapter 16.

[5] See November 1994 issue of *Morenci Copper Review* for pictures and story of the incline and its remains as explored by Ted Cogut and John Korolsky.

The Metcalf, Wilson and King inclines

The Metcalf mines of the Arizona Copper Company were located on the east side of Chase Creek on a mountainside overlooking the town of Metcalf. They were mostly shallow underground mines producing copper oxide, carbonate and silicate ore. The mines were serviced by two inclines, the Metcalf and the Wilson. These were both about 500 feet long and dropped ore down from the mines to storage bins, then onto the Coronado Railroad, over which the ore was hauled to the Arizona Copper Company mill and smelter in Clifton.

A tunnel was driven through the mountain from the Metcalf Mine workings, connecting with a towering wooden trestle which crossed King Canyon, a side canyon to Chase Creek Canyon. The south end of the trestle connected to the King incline, which traversed several hundred feet up King Ridge to the King and North King workings, including the Jamieson pit. The trestle and tunnel were used to transport the ore from these mines to the Coronado Railroad. There was also an inclined rail which dropped from the King side of the trestle and ran under the trestle to another tunnel that was access to some of the deeper mines on the Metcalf side. Although mining at Metcalf ended in 1919, the old trestle was still there during the early 1950s. Many times in the '50s we crossed under it doing geology and survey work and even walked across it a few times although it was pretty rickety. Sadly, it was knocked down one day with a drill road construction blast. All the old mine workings and remains of the Metcalf and Wilson inclines disappeared when operations began in the open pit of the Metcalf Mine in 1969.

The Queen incline

Not much is known about the Queen Mine and its incline. It was an underground mine and was located on what later was called the Queen Hill section of the Morenci open pit. The mine and incline were located on the west side of Chase Creek. The incline lowered ore from the mine to Chase Creek Canyon where the ore, as with the other inclines, was loaded onto cars on the Coronado Railroad. It was then hauled to the Arizona Copper Company mill and smelter in Clifton.

The Eagle and Fair Play inclines

Frank Ruedas was born in Morenci in 1906 and lived there until his retirement from the Phelps Dodge Morenci Branch in 1969, when he moved to Safford where he now lives. He told me during an interview that there were two inclines in old Morenci during the days of underground mining.

One was the Eagle incline located near the Joy shaft. It was used to transport ore by little mine cars attached to cables from an early level of the Humboldt Mine on Copper Mountain down to the little railroad which serviced the Joy shaft and main drift of the Humboldt Mine. Frank tells us the railroad was electric powered by trolley wires and connected with the Longfellow incline and with the No. 6 concentrator through two tunnels.

The Fair Play incline was also located in Old Morenci. It extended from a drift high on Copper Mountain, probably part of the Fair Play claim, down to another baby gauge railroad which ran through town on the road behind the Phelps Dodge mercantile store. The trains were pulled by little steam locomotives. The Fair Play incline dropped through the place later occupied by the Morenci hospital. [6]

The end of the inclines

As each of the underground mines were abandoned in the 1920s and up to 1932, the inclines in the district were no longer needed. Their rails were salvaged for scrap, leaving only the paths they followed as reminders of their roles in early-day mining. At the time of this writing, the only evidence of their existence is a part of the Shannon incline and also a part of the Coronado incline, both of which can still be seen north of the open pit mining operations.

[6] Frank Ruedas, personal conversation, July 1997.

An early photo of the Longfellow incline, circa 1890. The Coronado Railroad appears at the bottom, but the Shannon Railroad, which would run along slightly above the Coronado, was not yet built. Also not yet built were the ore bins that would later appear at the foot of the incline beside the Coronado Railroad. *Courtesy Roberta Troxell.*

Top of the Longfellow incline, circa 1900. The ore cars will be lowered down the incline to ore bins in Chase Creek Canyon.
Courtesy Phelps Dodge Corporation.

In this photo taken in about World War I, miners pose beside a cabled ore car at the base of the Metcalf and Wilson inclines. The Metcalf incline is on the left. *Courtesy Roberta Troxell.*

Ore cars loaded with ore from the Shannon Copper Company mines located near the top of this incline were lowered down to the bins where the ore was loaded onto the Coronado Railroad train. The Shannon Company later built its own railroad. *Courtesy Phelps Dodge Corporation.*

The Metcalf (left) and Wilson inclines in about 1917. They were used to lower ore from the Metcalf mines to the storage bins and the Coronado Railroad. *Courtesy Phelps Dodge Corporation.*

Workers pose at an ore car at the bottom of the Coronado incline. At 3200 feet, it was said to be the longest in the country. It rose 1200 feet in elevation. *Courtesy Roberta Troxell.*

Looking down from the top of the Coronado incline in about 1915. Far below in the middle distance is the town of Metcalf.

Courtesy Phelps Dodge Corporation.

Photographed in 1986 at the top of the Coronado incline where it last had been used 63 years earlier, this tugger was put on display in the Morenci plaza in 1990. *Photo by Ted Cogut.*

This photo, taken in 1986 at the top of the Coronado incline, shows the cable reel and part of the steam-powered machinery that had lowered and returned ore cars on the steep mountain side for the last time in 1923.

Photo by Ted Cogut.

A view in 1921 of the base of the Longfellow incline, Coronado Railroad, and the Clifton-Metcalf highway which later became the southern part of the Coronado Trail. *From the Merna Robison collection.*

11

Frank Ruedas worked at Phelps Dodge Morenci from 1925 to 1969....WCC

Our knowledge of Old Morenci would not be complete without the humor and recall skills of Frank Ruedas, who was born in Old Morenci on May 27, 1906, the son of Jesus Pedraza and Guadalupe Candalaria Ruedas.[1] This was the time when Old Morenci was the company town of the Detroit Copper Mining Company and the Arizona Copper Company. Detroit Copper Mining Company was by then owned and operated by Phelps Dodge. Frank's dad was a miner in the Arizona Copper Company's Humboldt underground mine.

Frank was born in a house on a hill overlooking the Joy shaft head frame, which was the main shaft for the Arizona Copper Company's Humboldt Mine. His grade school was the Longfellow School, located next to the Morenci Hotel.

One of Frank's early memories was the 1915 strike against the three major copper companies — the Detroit Copper Mining Company, Arizona Copper Company and the Shannon Copper Company. Frank's dad moved his family to Duncan because of the threat of violence by strikers against company sympathizers. They lived in a "bunker," a cell-like hut. Some of the earlier families to move got to live in tents, which provided better living quarters than the bunkers. It wasn't until after the strike was settled that the Ruedas were able to move back to their home in Morenci.

When still a young boy, Frank delivered lunches from a local boarding house to men working on Burro Alley, the main road in Morenci. He remembers the Longfellow hospital, which was located on one of the upper levels of town. There was also another hospital where the teachers' dormitory was later built.

---

[1] Conversations with Frank Ruedas, July 1997.

There was a Morenci "Chinatown" in the old days. Frank remembers seeing Chinese men carrying big baskets of vegetables on their shoulders. Chinese immigrants first came to the area in the 1870s when Henry Lesinsky employed them to work in the Longfellow Mine.

Mules were kept underground for years, laboring in the mines. Frank remembers them finally being brought out of the 5th level of the Humboldt Mine blindfolded, so the light wouldn't blind them. They were then kept in a corral in the Fair Play section of Morenci. The ones that died were buried in Fair Play.

Violence often ran high in Morenci. Frank recalls several bombings of the Catholic church. The Indian wars were over, but the lawless element was there to be reckoned with.

Blue Ribbon Beer was the most popular brand in town. The bottles had real blue ribbons and were prized collection items for Frank and his buddies.

The Morenci Hotel, built in 1901, had a restaurant on the ground floor. In later years, the Morenci Water and Electric Company and the rental department offices were located on that floor.

Underground copper mines were around and underneath Morenci. There were two inclines in Morenci on which mine cars ran up and down the mountainside on rails carrying ore and supplies to and from mine entrances. The Eagle incline served one of the levels of the Humboldt Mine and was located behind the Joy shaft. The Fair Play incline served the Arizona Central Mine.

When Frank was a kid, there were no automobiles in old Morenci. But Morenci had trolley cars. The trolley cars were surface mine cars which ran by connecting with overhead electric wires from the Humboldt Mine in Morenci through a tunnel to the old Longfellow Mine. Another trolley line ran from the Humboldt Mine through a tunnel to the No. 6 concentrator.

Some of the old underground mines provided relief from the hot summer sun. Frank remembers a big room, or stope, in the Lone Star tunnel where people went during the summer to stay cool. That old mine is now part of the Southside Extension presently being mined by open pit methods.

Jessup's Dairy was located on the road to Newtown, a settlement that was a short distance south of Morenci and south of the Detroit Copper smelter slag dump. The dairy delivered milk every morning by horse-drawn cart. Later on, a dance hall was located where the dairy had been.

Billiards was a popular source of entertainment. Jack Cox owned and operated the pool hall. Even some of the ladies from a certain house in Newtown, who provided their own brand of entertainment, would show up to play billiards.[2]

Baseball was a very big sport in Morenci in the early days. At least two players went on to distinguished careers in the major leagues. Phelps Dodge supported a team, and while they were boys, Frank and his friend, Frank Alvidrez, were the team's scorekeepers. They traveled with the local teams to other towns in Arizona and New Mexico. Frank remembers Alexander Peete, who played for the Phoenix Giants. One time, they played in Morenci, and Alexander stayed and went to work for the company. The Peete family has been in Morenci ever since. The keen interest the major copper companies had in sponsoring baseball teams began in the early 1900s and lasted into the 1930s.

Another sport that was popular when Frank was a youngster was rebote, a type of handball. It was played on a three-cornered court which meant the ball, made of hard rubber, could carom off the wall and come at you from practically any direction. There was a line two or three feet high on the wall. The ball had to hit above that line. There were rebote courts in Morenci and in Metcalf. The Morenci rebote court was located by the Empire Theater on AC Hill. The hills of Morenci were designated "A", "AC", "B", "C" and "D".

Later on, softball, bowling and tennis were also popular. Some played golf. The golf course was located on the tailing dam at Bunkers created by the No. 6 concentrator. The area named Bunkers later became the site of the new Phelps Dodge mill and smelter, built during World War II. Frank was one of the honored ones to be invited to play golf there with football coach Frank Simoneck.

A particular brand of softball was played on the tennis courts. This involved a ball so huge that it wouldn't travel very far no matter how hard it was hit. Regular softball was played in the plaza below the PD Store. Frank played in the outfield, and that's how he found out he needed glasses.

Soccer (English football) was played on what would become the football field. Frank remembers a Mexican man kicking the ball out of the park. When the ball was finally retrieved, it was completely torn up. Another of the

---

[2] They came from one of those establishments, the "house of ill repute," that could in those days be found in many communities across the country. Old timers recall the one in Newtown as "The Green House." Business was good enough, it was said, for the madam to buy a new car every year.

early athletic pastimes was roller skating. There was a large, impressive-looking building which housed a skating rink located at the bottom of the steep road leading to Eagle Creek.

When Frank was in high school, one of his athletic talents was running. There used to be a relay race run from Clifton to Morenci. Each runner ran 1/4 mile. Frank was assigned to run up the 10% grade, the toughest hill on the course, he said, " because I was the skinniest guy on the team." He was also on the tennis and basketball teams. Frank continued playing tennis into his adult years. He said the only person he couldn't beat was Warren Fenzi, Morenci Branch general superintendent, who was very tall and athletic and could reach all over the court. Fenzi went on to become president of Phelps Dodge Corporation.

A Most Unusual Football Field

The Detroit Copper slag dump later became the location of the famous Morenci High School football field where attempts at field goals could only be kicked toward the east side of the field. To kick toward the west side would have allowed the ball to go bouncing off the cliff and into Morenci Canyon. When the opportunity for a field goal arose, the teams often had to switch sides in order to be kicking toward the east.

Frank played basketball in his younger days. Once, when his team was playing Clifton, Frank had a jump ball against Jesus Zorilla. Frank figured he had it made because Jesus was so short. Jesus, however, had a scheme to compensate for his lack of height. As they went up for the ball, Jesus kneed Frank in the crotch! To add insult to injury, Jesus tipped the ball to a teammate, and no foul was called!

In those days, playing courts weren't exactly built according to regulation. The basketball court in Clifton was so short, the free throw courts intersected! Rafters overhung the basketball court in Duncan. Duncan boys knew how to lob over them to score. Morenci boys kept hitting the rafters.

While he was a student, Frank worked part time in the Phelps Dodge mercantile store. Frank insists that, contrary to some historical articles, there was no mortuary in the store. They did sell coffins in the furniture department, where he worked. They also sold 50-gallon wine containers in the furniture department. It was not a train station or post office, as has been written by some historians. These were located nearby, the Morenci Southern Railroad station being located just below the store in what later became a parking lot.

September 16, Mexican Independence Day, was always celebrated in Morenci. Frank Ayer, Phelps Dodge manager during the 1920s, would always have bleachers and a stage for the chorus girls set up in the town square for all the festivities

In 1924 or 1925, copper was made from an early Solution Extraction-Electrowinning process

Also while he was a student, Frank worked in a test plant at the No.6 concentrator. The plant was a forerunner to the solution extraction-electrowinning process we have today. Pete Gomez describes the plant in Chapter 17, featuring his recollections. Frank recalls that the plant was started in 1924 or 1925. He remembers using a hammer and chisel to scrape copper off the lead plates. He also ran the crane which set the lead plates into the electrolyte.

Frank graduated from Morenci High School in 1926. That high school building was built in 1921 and torn down in 1950. It was located beside the Morenci Club. After graduation, Frank was offered a transfer to the Phelps Dodge mercantile store in Douglas, Arizona. He had already bought a suitcase when he heard of an opening in the PD accounting department in Morenci. He went to Harry Hill, chief accountant, asked for the job, got it, and "was there until I retired." He started in "addressograph," printing lists, payroll checks and statements. He also helped the payroll clerk and was timekeeper for the Humboldt Mine. Timecards were posted at 6:30 every morning.

Hard times fell on Morenci after the stock market crash of 1929, precipitating the Great Depression. The mine tried to struggle on and keep operating, but in 1932 Phelps Dodge was forced to shut down the mining operation and everyone was laid off. The skills Frank had developed in accounting helped him get a government job as timekeeper for the ERA and WPA. Frank said he had it good, making $125 a month. He even bought himself a Model A Ford.[3]

A beautiful Morenci girl, Elvira Arrieta, was very popular. In fact, in a combined Morenci-Clifton vote, she was voted most popular in Morenci.

[3] The ERA, Economic Recovery Administration, and the WPA, Works Projects Administration, were two of President Franklin Roosevelt's efforts to cope with the harsh economic realities of the Great Depression. Laborers on WPA projects only made $39 a month. During the depths of the Great Depression, a Model A Ford, which had sold for about $600 when new, could be purchased for $50 to $75.

Frank remembers the Zorilla boys in Clifton, who later owned and operated a grocery store in Clifton for many years, voted for Elvira. Frank won Elvira's heart and, in 1940, they were married. After the wedding, Frank Alvidrez and Frank Martin drove them to Lordsburg, where they caught a train to El Paso, then another train to Mexico City where the happy couple spent their honeymoon.

Frank and Elvira had four children: Frank A., the twins Hector and Victor, and Daniel. The Ruedas family house was located on "C" Hill next to the old road to Eagle Creek.

In 1942, Frank was recalled to the Phelps Dodge accounting department. This was the year the open pit mine delivered the first ore to the new mill and smelter. In 1953, Frank was promoted to head timekeeper, a job he held until his retirement.

While talking about those days, Frank remembered a young man who worked at the rental department. Whenever finances were being discussed the young man liked to say: "money is no object." But money, apparently, *was* the object of his affection. One memorable morning the young man didn't show up for work. He was gone, never to be heard of again, and so was all the rental department cash.

Frank retired from Phelps Dodge on July 1, 1969 after 44 years of service. He had been head timekeeper for 16 years. In 1969, he bought his retirement home in Safford and has lived there ever since. He has remained very active, even in retirement. Some of his activities include: vice president of the Civitan Club in Safford; Meals on Wheels — he picked up the meals from the American Legion and delivered them to the shut-ins; and he has remained active in the Knights of Columbus. [4]

I was amazed at and impressed with Frank when I interviewed him. At age 91, he is very sharp, funny, enjoys life and has a remarkable memory. He visits the people at Art's barber shop in Safford every day that they're open. They look forward to his visits and check on him if for some reason he doesn't show up. During his lifetime, he has made many friends in Safford, Morenci and Clifton and has set a great example for all of us.

---

[4] *Who's Who in Arizona* of 1958 included Frank as one of its outstanding people. The article listed these accomplishments:
Knights of Columbus, 4th Degree, Past Faithful Navigator (with medal); Boy Scouts of America District Chairman, Camping Activities Chairman; Greenlee County Tennis Club (an excellent player); Morenci Young Democrats, Charter President; Woodman of the World, 1954 Award; Morenci Lions Club (Tail Twister); Certified First Aid Instructor, 1930; Phelps Dodge Head Timekeeper; Democratic Convention Delegate; National Rifle Association; Coronado Gun Club; Latin American Club; and Regis Dance Club.

Frank Ruedas, the second generation of a four-generation Phelps Dodge family, holds a portrait of his father who retired from Phelps Dodge Morenci Branch in 1920. *Courtesy Frank Ruedas.*

Narrow gauge passenger cars at the Plaza in Old Morenci in the early 1920s. The large building at the extreme right is the DC store. The Morenci Hotel is left of the store. The Morenci Club is in the background at left. *Courtesy Frank Ruedas.*

The Joy shaft, shown in the center of the picture, was the main shaft of the Humboldt Mine, the first sulfide mine as well as the most productive of the district's underground mines. Mining began in the Humboldt in the 1890s and ended in 1932. This site was within the town of Old Morenci.
Courtesy Phelps Dodge Corporation.

12

The town of Metcalf flourished and vanished....WCC

Almost all of us have heard about the old mining town of Metcalf, it was said to have once been the biggest town in Arizona, but most of us were not around when it was in existence. The town was located in Chase Creek Canyon in the area a little north of the Morenci open pit and now occupied by the company's heavy duty truck garage, the old Metcalf primary crusher and the adjacent old intermediate ore storage facility. Metcalf was a pleasant mountain town where everyone had a view.[1] How and when did the town come into existence, and what led to its demise?

Robert and James Metcalfe located mining claims in 1872 near this location. By 1875, Robert Metcalfe had established a camp close to his mines. This camp, which later became the town of Metcalf, (with the trailing "e" in "Metcalfe" dropped) was in an isolated and picturesque area near the confluence of Chase Creek and Santa Rosa Canyon. About 50 people called it home during the 1880s. In 1882, the Arizona Copper Company bought up all the claims in and around Metcalf that then became its company town.

By 1881, the Longfellow Copper Company had built the Coronado Railroad up Chase Creek from Clifton to Metcalf. This was a "baby gauge" railroad served by tiny locomotives similar to the little locomotives on display in Morenci and Clifton. With this increased accessibility mining activity flourished and the town grew. At that time, there were several active copper mines near Metcalf. One of these, which provided ore for the Arizona Copper Company and later for Phelps Dodge, was the Coronado Mine, located high in the mountains above and west of the camp. A railroad was built to extend up Santa Rosa canyon to connect the Coronado Railroad to the foot of the towering incline which, at its top, was in turn connected to still another railroad, this one only about a mile long that took a horseshoe route to the mine site.

[1] Emma Ruiz Pando. See Chapter 27 for more about the town where she lived for many years.

The Arizona Copper Company also actively mined the hills directly overlooking Metcalf on the east side of Chase Creek, in the area which later became the open pit of the Metcalf Mine. Two steep rail inclines dropped ore from the mines down to Metcalf into the Coronado Railroad trains that hauled the ore to the smelter in Clifton. The Coronado Railroad offered regular passenger service as well as copper ore and other freight haulage between Metcalf and Clifton.[2]

The Shannon Copper Company, organized in November 1899,[3] was another of the major copper companies in the district. Startup of the Shannon Copper Company, around 1900, was very beneficial for the town of Metcalf. The Shannon mines were located directly above the Arizona Copper Company's Metcalf mines which were serviced by the Wilson and Metcalf inclines. Shannon's mines, composed of shallow drifts, stopes, shafts and adits in the top of Shannon Ridge, were serviced by the Shannon incline. The mines of both companies towered above the town of Metcalf. The copper minerals mined were oxide copper ores such as azurite, malachite and cuprite. Excavated material from these workings is still visible today.

The first school in Metcalf was built in 1895. By the turn of the century, Metcalf had become a bustling community of miners and their families. Arizona Copper built a hospital there in 1908. By 1910, Metcalf boasted a population of about 3,000. In addition to the school, it had a railroad and a post office. The first automobile drove into town in 1913. The State Bank of Metcalfe opened in October 1915.[4]

Sometime after 1903, a road was built connecting Clifton and Metcalf, paralleling the railroad in Chase Creek Canyon. Al Fernandez, a Clifton resident and well-known area historian, traveled with his dad to Metcalf on a regular basis from 1914 to 1935, hauling groceries and feed for the mules that were used to carry ore and firewood. The two first hauled supplies in a horse-drawn "express wagon" and later, in the days before pneumatic tires became common, in an International pickup with hard-rubber tires.

The 3-B store was an establishment well-remembered by a number of area old-timers. It was located in Metcalf and was owned by Bianes and Valdez. The three B's stood for Bueno, Barato y Bonito (good, cheap and pretty). The basement was a warehouse, the bottom floor contained general merchandise, the top, a saloon. The Fernandez store sold them groceries.

---

[2] We are indebted to Al Fernandez and the late Frank Alvidrez for much of the information about the town of Metcalf that came from extended conversations with them in 1997.

[3] Norman Carmichael and John Kiddie, *Transactions, American Institute of Mining and Metallurgical Engineers, Vol.LXX, 1924, "Development of Mine Transportation in Clifton-Morenci District," 832.*

[4] James S. Patton, "Farewell to Metcalfe," Installments 8,14, *Copper Era,* 1940.

The 3-B store had a bank loan for $10,000 to cash checks for the Arizona Copper and Shannon Copper miners.

In 1916, the famous Metcalf High School was completed in Santa Rosa Canyon. Students in the elementary school waited in eager anticipation for their graduation so that they could transfer to that state-of-the-art educational facility.

Employment was high at the Arizona Copper and Shannon mines during the first two decades of the 20th century. Smaller companies, such as the King, Standard and Queen mines, also prospered. The town of Metcalf continued to grow. There were two company stores, one for each company.

The fortunes of the Shannon Company followed those of the Arizona Copper Company. It prospered, but could have done even better, especially during World War I, if it hadn't been for two long and violent strikes in 1915 and 1917 which shut all the companies in the district down for extended periods. (These labor problems are discussed in a later chapter.) The end of the war was followed by a severe depression in the price of copper. Because of the low price and also because the copper content of the remaining ore was too low, Shannon could no longer produce copper at a profit.

Today, one can still see abundant evidence of the old Shannon operations. Parts of the Shannon incline still exist, including skeletal remains of the incline superstructure and heavy-duty machinery parts which powered the hoist. Along with this, the old workings of the mines remain at the top of Shannon Ridge above the site of Metcalf which is now occupied by the present mine shops of the Morenci and Metcalf open pit mines. Slag from the Shannon smelter can still be seen near Shannon Hill in Clifton. The old mine workings and machinery give silent evidence of a booming mining company that suddenly ceased to exist in 1919.

By 1919, all the mines that had supported Metcalf were shut down. In that year, Arizona Copper bought all the holdings of Shannon Copper, including the mines, railroad and concentrator. Two years later, copper had dropped to 11 cents per pound from a price of 33 cents a few years before and Arizona Copper was unable to continue operations. It sold all its holdings to Phelps Dodge Corporation in 1921. This brought some renewed life to Metcalf. In 1922 Phelps Dodge reopened the Coronado Mine which became Phelps Dodge's second highest producing mine. However, that operation only lasted about a year. The Coronado Mine was shut down permanently in 1923, ending any hope of survival for Metcalf.

Many good things come to an end, and this was the case with our town of Metcalf. Residents began leaving Metcalf and by 1930 only about 500 inhabitants remained. The high school continued in use on a limited basis until the last families left Metcalf in the late 1930s. The school was sold to a scrap dealer for $1,000. The town was gradually dismantled as residents moved out. Some of us who lived in Morenci in the 1950s and 1960s recall the walls of an old building that were still standing in Metcalf. Those ruins were the remains of the "3-B" store built in the early 1900s — the only remnants of the town, except for evidence of a few building foundations.

Although the town is gone these many years, the name "Metcalf" remains. Phelps Dodge gave the name to the open pit it began mining in the Metcalf area in 1970 and to its second concentrator built in 1974.

The Clifton-Morenci-Metcalf area was a rough place in the late 1800s and early 1900s. Note written on back of this photo reads: "Metcalf bank vault — robbed by lone man."
From the Merna Robison collection.

The Shannon Railroad drops down to the town of Metcalf in the right background. Circa 1915. *Courtesy Roberta Troxell.*

The hard-working burros in Metcalf carried big loads.
From the Merna Robison collection.

Metcalf, Arizona in about 1920. The large, light-colored building on the left is the Metcalf High School. A trestle on the Shannon Railroad crosses Santa Rosa Canyon in the center of town. *Courtesy Roberta Troxell.*

A cowboy who rode into Metcalf on the dark horse provides milk to a little girl from milk cans strapped to his pack horse. *Courtesy Roberta Troxell.*

A rancher with a large load of provisions passes through Metcalf. *Courtesy Roberta Troxell.*

At the base of one of the Metcalf Mine inclines, mule-drawn ore cars are loaded from bins.
Courtesy Phelps Dodge Corporation.

The headframe and steam power plant of the Matilda shaft of the Coronado Mine. The mine was permanently closed in 1923.
Courtesy Phelps Dodge Corporation.

Recognizable by its flat roof as one of the later models of the baby gauge locomotives, this little engine pulls a load of ore from the mines at Metcalf or Shannon along the Coronado Railroad, heading for the smelter in Clifton. *Courtesy Roberta Troxell.*

With all the other town buildings already gone, the magnificent Metcalf High School stands all alone, waiting for demolition. It was built in 1917 and abandoned in just three years when the depressed copper market forced closure of the mines. *Courtesy Roberta Troxell.*

Metcalf's 3-b store was the last building in town to be torn down. The three B's stood for "bueno, barato y bonito" — "good, cheap and pretty."

Courtesy Phelps Dodge Corporation.

The town of Shannon, early 1900s. *From the Merna Robison collection.*

13

Al Fernandez has lived much of the history of Clifton, Morenci and Metcalf.... WCC

Because he has lived most of it, no one knows more about our local history than Al Fernandez. Al has lived in Arizona since 1912, in Clifton since 1914, and has been very active in the Greenlee County Historical Society ever since its inception. He has given many lectures, conducted hundreds of tours of Chase Creek (the old Clifton business district) and has written many, many articles which appeared weekly in the *Copper Era.* Ted Cogut and I have interviewed Al, listening to fascinating accounts of his life and of the early days. We're grateful to him for spending a lot of time with us, helping us to see what life was like in those years now long past.[1]

Al was born in Mexico in 1901. His family was forced to leave their home because of threats from Pancho Villa and his army of 600 men who raided their home and homes of their neighbors. For a while, Al and his family couldn't get out of Mexico. Then little Al helped his father bribe the conductor of a train carrying animals and 1,000 people to El Paso. They then were allowed to board the train. When they arrived in El Paso, the conductor recommended Solomonville, Arizona as a good place to live. They got train tickets and arrived in Solomonville, which was called "Pueblo Viejo." The year was 1912 and Al was 11 years old.

They happened to arrive on Halloween night and received a rude welcome. While on their way to their new home, they ran into a wire that pranksters had strung between trees which knocked them out of their wagon!

Upon their arrival, they met I. E. Solomon, after whom the town was named. Solomon is prominent in our history. He supplied all the mesquite as fuel for the early copper smelters located along Chase Creek. The town of Solomonville was established as a source of the mesquite.[2] Mr. Freudenthal

---

[1] From conversations with Al Fernandez, 1997. His articles were titled "County Chronicles."

[2] James Colquhoun, in his *Early History of the Clifton-Morenci District* adds that Solomon used the land cleared of mesquite along the Gila River to produce "food supplies for man and beast," William Clowes and Sons, London and Beccles, c1920, 44.

ran the town. He was related to Julius Freudenthal, who was one of the owners of the old 1870s Longfellow Copper Company, near the site that became the town of Morenci.

Solomon and Freudenthal helped Al and his dad start their chicken farm. They also had ducks, geese, guineas, a cow, two horses and a wagon. They transported eggs to Clifton on the old toll road which ran through the Black Hills in the same canyons now traversed by Highway 191. There was a toll house on one of the hills between Solomonville and Clifton. The little settlement of Coronado was located at the point where the road reached the Gila River. This was near the present location of the Gila River bridge. The Arizona and New Mexico Railway, which is now the Southern Pacific connecting Clifton and Lordsburg, had a platform at Coronado.

Al and his dad made their first trip to Clifton from Solomonville in 1912. The road was very steep. Wagons with a ton of hay had to rent four fresh horses from the man at the toll house. The toll was 10 cents per wagon and 10 cents per horse. Al and his dad reached the Gila River at Coronado about 4 p.m. There had been a lot of rain and the water level too high for a crossing. All the travelers with wagons waited two days for the water to go down. Then everyone decided that young, healthy Al could wade across to check the depth. He made it across with the water up to his chest. This gave the wagoneers heart, and Al's wagon and four others made it across safely. Al and his dad arrived in Clifton that afternoon and slept in their wagon that night.

Al knew some English words when he started school at age 11. He was the biggest kid in kindergarten! As soon as he learned a little more English, he was transferred to the third grade.

Al's main morning job at that time was to milk the cow and give a quart of milk to Mr. Freudenthal. Then he would load eggs into the incubator in their front room. He gave the eggs one half-turn every morning. They were heated by kerosene lamps and took 21 days to hatch. School kids would come to see them hatch. Another chore was to churn butter and take it to Mr. Freudenthal's store for sale.

Bill Clark was one of only three car owners in Solomonville. He was the stage driver. For a fifty-cent fare, Clark would take a family to Safford and back.

One day, Al and his dad saw a long box being unloaded at the store. It contained six car horns, a big shipment considering there were only three cars in Solomonville. Al's dad bought one of the horns and installed it on his horse-drawn express wagon.

Driving a horse-drawn wagon on the road between Solomonville and Safford in those days could become an experience to be remembered. There were two deep ruts in that road. One day, while Al was driving the horse and express wagon to Safford, he caught up to a horse-drawn rack in the ruts ahead. Al blew the horn his dad had just bought, and the man driving the rack maneuvered his wheels out of the ruts because he thought a car was coming from behind. Al says: "When he saw me — a kid — he got mad!"

They move to Clifton in 1914

An irrigation ditch in Solomonville flooded, drowning all the Fernandez family's chickens. They lost everything. This was in 1914. The whole family of six then pulled up stakes and moved to Clifton.

Al's uncle worked in the mines in Morenci. He put up the money, Al's dad supplied the team of horses, and they started a grocery store. They rented a house from Joe Villescas. The family prospered.

Chase Creek Street was the business section of Clifton. There were pool halls and saloons with rooms upstairs served by dumb waiters. Player pianos played all day and well into the nights. The Empire and Royal Theaters were on Chase Creek. The admission was a nickel, later raised to a dime. On Clifton's east side there were two banks, two hotels, two restaurants and a Chinese laundry. From 1915 to 1923 the streets were cleaned with a plow pulled by two mules. Then a steam roller with a one-piston engine would roll in to smooth out the bumps.

When they got settled and had made a little money, Mr. Fernandez bought a truck from Kelley and Lines in Safford. This was only the third truck to arrive in Greenlee County. It had solid-rubber tires, a magneto and petcocks. In the winter, one had to pour gas between the petcocks to start it. The truck was the object of pride for the whole family.

The store and the house were located on Chase Creek in Clifton, the house upstairs from the store. They were the competition for the Arizona Copper Company store, which was located in the building now occupied by the American Legion in Clifton.

Al's dad had been buying "Old Homestead" flour from the ACC store to stock his store. But the management of the company store later decided they didn't want to be aiding the competition, so they cut them off, forcing the Fernandez store to buy "Diamond M" brand flour. The Fernandez store also stocked materials from Spain for the Spaniards who worked in the mines.

They delivered groceries to people in Morenci, Metcalf and to those living in the little town of Shannon, also called "Metcalf-Shannon." The deliveries to Metcalf were made on Tuesdays and Saturdays. At first, this was with one-horse express wagons because the trail was too rough for the truck and later with the hard-rubber-tired truck.

The town of Shannon was located at the top of the Shannon incline above Chase Creek, north and east of Metcalf. The boarding house there bought eggs, milk, bread and garbanzo beans. One day, the Fernandez family asked the Shannon Copper Company manager if they could bring their groceries up the Shannon incline in an ore car. In a very courteous manner they were told others had made the same request and were turned down. This meant the groceries would have to be hauled some other way up that long, steep, tortuous trail from the bottom of Chase Creek Canyon to the town at the top of the mountain. Four burros supplied by the Loya family, who had 30 of the animals, faithfully did the job.

Al not only remembers Metcalf's 3-B store described in the last chapter, but also has a clear memory of where almost every shop and home was located in Metcalf. The Arizona Copper and Shannon Copper companies both had company stores there. Three or four mules tied together would make deliveries to the stores. Al talked about the other businesses — a barbershop, tailor, bakery, pool hall, boarding house, bank, taxi service, candy store and seamstress, just to name a few. There were also two churches — Catholic and Presbyterian. Burros were the main hauling source, carrying ore, wood, groceries, furniture, whatever needed transporting.

Al mentioned the movie theater in Metcalf, owned by Frank Martin. He also owned two theaters in Clifton and one in Morenci. Even as late as the 1950s there was a Martin Theater in Clifton.

When Al was a kid, Metcalf was a booming mining town of about 4,000 inhabitants, about 2/3 from Mexico. The rest were Spaniards, Italians and American-born. The town deputy was named Farnsworth and was known by many as "Fonzo."[3] Alex Arnett and Jose Maria Sanchez were also deputies. Jose was also the pool hall proprietor.

---

[3] James S. Patton, "Farewell to Metcalfe," Installment 11, *Copper Era,* 1940. Patton wrote: "C.H. Farnsworth had been a veteran of the Spanish-American War, an Arizona Ranger, and a prison guard. He came to Metcalfe in the early 1900s and served for many years as townsite agent....He was a deputy sheriff....He was also a game warden and a United States marshal."

Baby gauge locomotives ran along the Coronado Railroad which followed the Chase Creek stream bed. The railroad connected Clifton and Metcalf. Besides the little locomotives, Al recalls there were three coaches and 12 ore cars. Ore from the Metcalf and Coronado mines was hauled in the ore cars to the bins at the Arizona Copper Company concentrator and smelter in Clifton. There was also regular passenger service. The seven-mile train ride between Metcalf and Clifton took about 40 minutes. The Shannon Railroad also connected Clifton and Metcalf and also ran along Chase Creek paralleling the Coronado Railroad. Ore from the Shannon Mine was delivered to Shannon's mill and smelter on Shannon Hill in Clifton. There was no regular passenger service on the Shannon Railroad.

Al recalls that Norman Carmichael was president of the Arizona Copper Company and that he paid for the Clifton School band uniforms. Al's 5th grade teacher was Miss Helen Fisher at the Colquhoun School in Clifton, named after James Colquhoun, former president of the Arizona Copper Company. She told her students that Mr. Colquhoun sent money for candy for the children every Christmas (he continued doing it until his death in 1954 at the age of 96).

James Colquhoun talked to Al's class once on a visit to Clifton. Al remembers that he was well dressed, intelligent and that he was a tall, slender man. He spoke of the history of Clifton. Colquhoun also talked about Russia. His son was in Siberia, and he had been there to visit him. He told the class Russia had unlimited resources. The country, he said, was a sleeping bear, and he was afraid it would awaken and replace us as the best nation in the world.

Hard times for the mining towns

In the 1930s, during the Great Depression, people moved from town to town, always hoping to find some way to make a living. All the copper companies in Arizona were struggling, including Phelps Dodge. The companies kept operating for a while, Al tells us, because the state gave them credit, then cut off credit, which was a factor as Phelps Dodge decided to shut down all mining in 1932. The companies fought the decision in court, but lost.

Phelps Dodge chartered trains to provide transportation for 800 or 900 families who wanted to go back to Mexico because there was no work in the mines. The Mexican government had a consulate office on Chase Creek to help with the process.

With industry and the economy at a standstill, the towns received no more taxes. In Clifton, city and county employees were paid with warrants (a promise to pay) because the governments had no money. Al's store took the warrants at par value. Many of Al's customers owed bills at his store.

Prices tumbled. Nothing was worth much anymore, not even your house. One man wanted to sell Al and his dad a five-room house. He wanted $100 — No. How about $50? — No. The man noticed a $15 guitar in the store (it only cost Al $11), he said he'd take it for the house. The deal was made and Al later tore the house down and sold the lumber for $70.

Meanwhile, word got around that gold could be found near Clifton and families began coming to town to try to scratch out a living, panning for gold. Many of them came to Al at the store for advice on where to pan. He sent some up the 'Frisco River, to Chase Creek and to Gold Gulch (between Morenci and Eagle Creek). He remembers 14 people panning for gold in the 'Frisco, five in Chase Creek and six in Gold Gulch.

Al had a license to weigh gold, and the gold panners would bring their flecks of gold to him for redemption. Gold was supposed to sell for $35 an ounce, but Al only got $26.50, and that's what he paid per ounce for redemption. As Al weighed their flecks of gold, the usual payout was anywhere between 5 and 35 cents.

One day, the Kinney family from Ada, Oklahoma came to town, asked Al where they might find gold, and began panning on the 'Frisco north of town. Three days later, Mr. Kinney came back with one pennyweight plus five grains of gold worth $1.45. While Al and Mr. Kinney were weighing the flecks, Mrs. Kinney walked around the store, picking up food items to buy. Not knowing that they had not exactly struck it rich, she had gathered up $5.45 worth of groceries. Al let them owe him the difference, starting the practice of charge accounts at his store. The Kinney family was only averaging one ounce of gold every two weeks, which wasn't enough to pay for their groceries. Al would let them charge the difference. After one year, the Kinneys owed Al $135.

Another gold-panning family, the Moricals, also owed money. Both the Kinneys and the Moricals came to Al one day and said they were going to leave, but would pay him when they got the money. After three years, the debts were still outstanding, so Al claimed the loss on his income tax.

One day, after another three years had passed, Mr. Morical returned to Clifton and dropped into Al's store. He had become wealthy, he said, and he paid Al the $85 he had owed. He then asked if Kinney had ever paid Al what he had owed and upon finding that he had not, he paid the $135 that

Kinney owed. Two years later, Kinney also returned. He too had become wealthy and approached Al to pay his bill. Al told him that Morical had paid it. True to their word, both men had returned to pay off their accounts.

The cattlemen would pay their bills every November. "Between the mining and the cattlemen we never lost a penny. These people were great," Al said. Al also sold to the farmers. A government plan called "Milling and Transient Bill" helped many of the local farmers during the Great Depression years.

The Fernandez store was in the Cascarreli building on Chase Creek. It's the one with the flag pole on top. The family lived upstairs. After Al's parents died, his sister went to California and Al continued to live there for a total of 55 years. With only $148, he started a new business, the ABC market on North Coronado Boulevard. He said he picked a good time, because that was about when the open pit Morenci Mine was started, around 1937. Al had five faithful employees, and the store enjoyed prosperity for many years.

Al's wife was Pilar, nicknamed "Picky." She was born in Clifton and was a Palicio. They lived in the beautiful Palicio house behind the present Greenlee County Historical Society museum on Chase Creek Street. Charles Spezia is now the proud owner who lives there. Al and Picky have a daughter, Ruth. Al's Picky passed away in 1987.

Today, Al lives in a beautiful house on Chase Creek and still gives detailed tours of the historic Chase Creek business district in Clifton. In 1993, Al wrote historical notes for Ted Cogut centering on the various Chase Creek businesses that he also describes during the walking tours. Here are just a few excerpts:

"In 1911 and '12 this city block (Chase Creek)," Al wrote, "of commercial business buildings was completely destroyed by fire. In 1913, they were rebuilt by using the walls that were still standing."

Gerardo Zorilla was the owner of a grocery store whose tenant was Chief Garcia. This was "where you got more Boston Baked Bean candy for a nickel than any other place in town."

Hipolito Cascarelli was the owner of a restaurant whose tenant was Jim Sing. The restaurant "employed about eight people. About 30 customers ate there at noon. They were employees of the smelter and the round house from across the creek."

The southeastern half of a building owned by the Spezias was the law office of Henry Kearney. Al tells of a humorous incident that occurred there:

"During prohibition, Mr. Seidel, the agent for the American Express Co., called me and said, 'Al, will you go across the street and tell Mr. Kearney that I have a case of law books that is consigned to him? And tell him to come and pick them up right away because the law books are leaking.'"

When he was a 5th grade student in the Colquhoun School, Al had an unforgettable experience. "In the lunch hour," Al wrote, "I used to run to our home in upper Chase Creek, hurriedly ate and ran back so as to get back in time for our afternoon session."

One day, while running home for lunch, Al caught up with W.C. Young, a 200-pound man with a crippled left arm, who had been made a deputy to help control acts of violence in Chase Creek. Mr. Young was attempting to climb the stairs leading to an upstairs law office. He wanted legal advice regarding a town ordinance that had just been passed which banned the keeping of livestock within the city limits. Mr. Young owned a livery stable with about 10 horses and the ordinance would ruin his business. The stairway railing was on the right side of the stairs. Al described what happened:

> In order to use the railing, he had to switch his cane from his good right arm to the left. He started wobbling, lost control of his equilibrium. So with an overloaded mind — the accumulation of problems, his physical condition, the passage of the ordinance that was going to ruin his business along with his inability to climb the stairs — he pulled his gun, pointed it to the temple of his head and fired. He dropped dead two feet in front of me. The acoustics of the hallway amplified the sound of the shot. Four people from the drugstore came out to see what happened. His body lay sprawled on the floor in a puddle of blood. Frightened, I tried to leave. I was told not to because I was a witness. I missed my lunch and half my class at school.

Al is not one to let things go. Recently, at the age of 95, he was repairing the ceiling of the kitchen in his beautiful house on Chase Creek when he fell, breaking a hip. It did not take him long to recover. Al has many friends and is always glad to have visitors. When I asked him how he was doing, he said, "I'm living the life of Riley."

Chase Creek Street in Clifton. A horse-drawn express wagon, such as would have been used by Al Fernandez and his father to deliver groceries, is parked on the left side of the street. Photo is believed to date to the early 1900s, possibly as late as the World War I era. *Courtesy Tom Rees.*

Store delivery service in the early 1900s — Morenci, Arizona Territory. The building is identified by a hand-written note as a lodging house. *From the Merna Robison collection.*

In about 1900, to help Morenci residents cope with the winter weather and to provide fuel for cooking, faithful burros delivered firewood. *From the Merna Robison collection.*

Built in 1897, the Arizona Copper Company power house still stands in Clifton.
Morenci Copper Review, November 1991.

The Chase Creek elementary school as it appeared during World War II.
Courtesy Phelps Dodge Corporation.

14

Labor strikes at Morenci in the early 1900s....WCC

Following an exhaustive research effort, in 1956 Roberta Watt (Troxell), wrote a detailed "History of Morenci, Arizona" as her Masters thesis at the University of Arizona. That work included a great deal of information on the Morenci strikes. I have borrowed heavily from her work for this description of the strikes of the early 1900s.[1]

Labor strikes plagued the copper companies and the towns of Clifton, Morenci and Metcalf during the early years of the 20th century. The strikes sometimes were violent and had devastating effects on the companies, communities and employees, including the managers and miners, and their families.

In 1903, miners of the Detroit Copper Mining Company (Phelps Dodge) and the Arizona Copper Company walked off the job demanding better working conditions and a 25-cents-a-day raise. Their leader was Jack Laustenneau, a Romanian, who was known as "Three Fingered Jack" or "Mocho" (crippled hand in Spanish). The strike lasted almost a month.

Morenci was then at its original location in Morenci Canyon, now the site of the open pit Morenci Mine; and it was in Graham County as Greenlee County had not yet been formed. The Graham County sheriff was Jim Parks and his Morenci Deputy was Dave Arzatte or "Little Dave."

With no settlement in sight, things got worse. It was believed Laustenneau would try to dynamite the Detroit Copper Mining Company plants. Ranchers, cowboys and employees were deputized.

The strikers met at the Longfellow lime quarry located on AC Hill in Morenci, overlooking the town. Speeches were made, energizing the crowd. Word got out that the strikers planned to hold Detroit Copper Company

---

[1] Roberta Watt Troxell, *History of Morenci, Arizona,* Master's thesis, 1956, 57-78

Superintendent Mills and the sheriff as hostages to force an audience for their demands. Company officials then appealed for aid to Acting Arizona Governor Isaac T. Stoddard.

Arizona Rangers and the National Guard were ordered to go to Morenci as was the U.S. Cavalry from Fort Huachuca and Fort Grant.[2] To protect them from the threat of being blown up, deputies guarded the looping trestles of the Morenci Southern Railroad.

Before the troops arrived, strikers from Morenci and Metcalf gathered in front of the Longfellow store. It seemed obvious they were threatening to destroy the store. When Parks and Arzatte walked into the crowd to try to calm them down, Arzatte was seized by the crowd. Parks pushed through the crowd to Laustenneau who agreed to disperse the crowd if Arizona Copper Company's Longfellow Mine Superintendent Paul Nicholas, who had taken a defensive position in a cave behind the Longfellow store, would talk to company officials to get some concessions for the strikers. Nicholas agreed, and the crowd dispersed, releasing Arzatte.

The same day, armed strikers gathered on Burro Alley in Morenci. It seemed a shooting episode was about to begin. Deputies took positions to protect the town. But before a shot could be fired, Morenci was flooded by a downpour. Water raged down Morenci Canyon, temporarily dampening the spirits of the would-be combatants, cooling the threat of a gunfight. During the 1960s, Head Sampler Vicente Rodrigues of the Morenci Mine's ore control division, who was three in 1903, still remembered the terror of the impending gunfight and the downpour which ended or at least postponed the confrontation.

When the Arizona Rangers arrived, they arrested Laustenneau. Other leaders of the strike also were arrested, ending the strike.

In 1915, another confrontation erupted as an attempt was made to organize the employees of the three major copper companies into a union. The union was the Western Federation of Miners. When Arizona Copper Company Manager Norman Carmichael, Detroit Copper Mining Company Manager Milton McLean and Shannon Copper Company Manager J. W. Bennie elected not to meet with representatives of the Western Federation, a strike was called. At this time, Morenci, Clifton and Metcalf were in Greenlee County, which had been created in 1909. The Sheriff was James Cash.

---

[2] Notes written by Merna Robison, a prominent resident of Old Morenci, state that the "Morenci cavalry," as she called them, stayed in town for six months.

The strike caused the shutdown of the power plant which plunged Morenci into darkness for three nights.

Charles Moyer, president of the Western Federation of Miners, arrived from Denver. The governor of Arizona, George P. Hunt, came to Morenci to listen to both sides. After the meetings, Hunt considered establishing martial law in the district. There were reports of damage as three thousand strikers paraded in Morenci.[3]

The three managers, fearing for their safety, decided to go to El Paso. They boarded a train from Clifton and headed for Lordsburg. When the strikers learned the managers had left, they swore out a warrant for their arrest, claiming the managers had incited a riot. The sheriff and his deputies confronted the managers in Lordsburg where a justice of the peace quickly dismissed the case. The managers proceeded to El Paso.

Governor Hunt ordered the National Guard to Clifton.[4] Many employees and their families who were loyal to the companies left town and established a camp in Duncan where they could live in safety. The companies now faced a problem. The deadline for completing the assessment work on all the mining claims in the district was fast approaching. Under the protection of 50 U. S. marshals, 486 workers from the camp in Duncan were brought in by the companies to do the assessment work.

The managers then agreed to return to Morenci to meet with five employees whom they had chosen from a list. A meeting was held, but no agreement was reached. The strikers then renounced their ties with the Western Federation of Miners and selected the Arizona State Federation of Labor to represent them.

In February 1916, the strike was settled with an agreement specifying that: (1) the companies will recognize no union, (2) there will be a sliding scale of wages from $2.50 to $5.31 per day based on the price of copper, and (3), a grievance committee will be established. Both the companies and the employees had been hurt by the strike — the company because of the loss of the good price of copper, and the employees because of the loss of wages

---

[3] In 1942, Merna Robison delivered "A Historical Talk for the Silver Anniversary Meeting of the Morenci Woman's Club." In a copy of that talk she wrote: "I do not remember its cause (the strike) but believe it was worked up by a few agitators with some connection with the enemy in Europe." (World War I was then raging in Europe.) She also wrote that: "The sheriff and governor provided so little protection that a thousand Morenci citizens took temporary refuge in Duncan, where they lived in a tent city, even though it was mid-winter, because they feared for their lives, and thought there was no safety in Morenci." The Ruedas family also went to Duncan. See Chapter 11.

[4] Tex Neel, a Morenci legend, was one of those arrested when the governor sent the National Guard in to regain law and order. See Chapter 22 for more about Tex Neel.

during the strike. World War I was continuing and the country was in dire need of copper. Copper's price was high, about 33 cents per pound.

In 1917, these new demands were presented to the three companies: (1) employee grievances would be taken up in regular form by the grievance committee, (2) an employee who refuses a contract would not be discharged or discriminated against, (3) seniority would rule in increases or decreases of the work force, (4) time-and-one-half would be paid for working on certain holidays and Sunday, and (5), adoption of the Miami wage scale. A rumor spread throughout the towns that the demands were influenced by the Industrial Workers Of The World (the "Wobblies"), a powerful union of national prominence.

Greenlee County Sheriff Arthur J. Slaughter stated he would favor neither side, enforce the law impartially and protect all citizens. John J. Donnelly, president of the Arizona State Federation of Labor, arrived in the district as soon as the list of demands was presented. The companies did not agree to the demands, and the strike was called on July 1, 1917.

As a precautionary measure, Sheriff Slaughter assigned 100 deputies to patrol Morenci, Metcalf and Clifton after the first week of the strike. Shortly after the strike began, a federal mediator arrived and started to confer with management and the employees. When he met with management, he said he felt the strike was not justified and that it had been influenced by a small number of men. The managers stayed out of town, returning only for conferences.

All efforts to settle the strike failed. In October 1917, a commission named by President Woodrow Wilson arrived to investigate conditions. The commission achieved a settlement under the following conditions:

1. Each company was to have its own grievance committee.
2. A United States arbitrator was to settle all disputes which could not be resolved by management and employees.
3. No man was to be prevented from working because he was not a member of the union.
4. Striking members were to return to work unless they had made unpatriotic statements about the United States.
5. Time-and-one-half would be paid for Sundays, overtime, Christmas and July 4th.
6. Wage disputes were to be settled by the United States arbitrator.

This agreement ended the third strike in the district and the companies were able to get back into production.

A troop of the Arizona National Guard rides through Morenci during the strike of 1903. Behind them, a passenger car of the Morenci Southern Railroad and a box car of the Arizona and New Mexico Railway rest on a siding. Both cars are narrow gauge. The large building on the upper right is the Morenci lodging house. *From the Merna Robison collection.*

Morenci as it looked in 1899, looking west. Detroit Copper Mining Company Manager Charles E. Mills' house is in left foreground. DCMC Superintendent Gordon McLean's house is in the center (with light and dark colored roof). Immediately behind and to the right, the building with three windows is the schoolhouse which also was used as a church. The large building standing alone in the far center is the new rooming house run by a Mr. Shell (sp?). The general office, surveyor's office and accounting department is at extreme left. At extreme right are a mine shaft and waste rock dump. A track leading from the shaft carried ore down to the ore chute where it was carried away by trains that ran on the track below. Note the water tanks and outhouses. A burro stands in the immediate foreground.

From the Merna Robison collection.

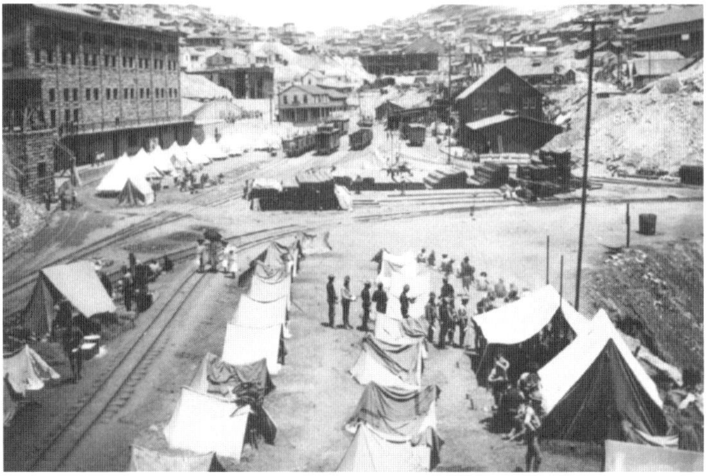

Arizona National Guard troops pitched their tents in the Morenci Plaza, 1903. The large building is the Detroit Copper Mining Company store.
From the Merna Robison collection.

Chow time at the mess tent by the Morenci Club, 1903.
From the Merna Robison collection.

15

Railroads made mining go.... WCC

The Clifton-Morenci-Metcalf area was still very much a part of the wild west in 1879 when the Longfellow Copper Company, under the leadership of Henry Lesinsky, built the first railroad in the district. The constant threat of Indian raids was one reason Lesinsky was anxious to build the railroad.

The Coronado Railroad

The Coronado Railroad, the first railroad in Clifton-Morenci, was what was called a "baby gauge" (20 inches between rails). It stretched from the foot of the Longfellow incline down Chase Creek Canyon to Clifton, a distance of about four miles.[1] At first there was no locomotive available. Mules supplied the power. They would pull the empty cars up the canyon from the smelter in Clifton to the ore bins at the base of the Longfellow incline. The cars would then be filled with ore from the Longfellow Mine, except for the last car. The mules and mule-skinner would climb aboard the last car, and they would all coast down Chase Creek Canyon to Clifton.

In 1880, the parts for a locomotive finally arrived. It was a tiny locomotive and was named "Little Emma." Local residents called it "El Vapor" because it made so much smoke starting its journey from Clifton to Longfellow.[2] Two locomotives very similar to "Emma" stand as historical displays today — one in Clifton and one in Morenci, memorials to the early days.

---

[1] James Colquhoun, *The History of the Clifton-Morenci Mining District,* London: John Murray, 1924, 8.

[2] Norman Carmichael and John Kiddie, *Transactions, American Institute of Mining and Metallurgical Engineers, Vol. LXX, 1924,* "Development of Mine Transportation in Clifton-Morenci District," 831; and William Ryder Ridgway, *Arizona Highways,* "Whistling Down the Canyon," July 1961.

The Coronado Railroad was extended up Chase Creek another three miles to the town of Metcalf in 1881 and thence another half mile to the foot of the Coronado incline. The incline was 1200 feet high and was very steep. At the top of the incline was another section of baby gauge railroad which extended about a mile west in very rugged terrain to the Coronado Mine. The Coronado ore body had been discovered by the Longfellow Copper Company and mining began there in 1881.

By 1882, Henry Lesinsky made a decision to sell his Longfellow Copper Company. The constant threat of Indian raids had to have been an important reason for his decision. He also knew a railroad out of Clifton was needed to more profitably send the ore to market and to simply connect Clifton with civilization. He made arrangements to sell all his properties, including the Longfellow Mine and the Coronado Mine, to the Arizona Copper Company of Edinburgh, Scotland. The town of Duncan, Arizona was named after J. Duncan Smith, one of the directors of the Arizona Copper Company.[3]

The Coronado Railroad ran a commuter train that carried workers to and from the smelter that Arizona Copper Company built in 1913 south of Clifton.

The Arizona and New Mexico Railway

After the Arizona Copper Company bought the Longfellow Copper Company in 1882, it immediately embarked on a three million dollar expansion plan. The expansion included building a railroad connecting Clifton to Lordsburg, New Mexico, 70 miles away, and other improvements.

Construction stopped on the railroad shortly after it started, not because of Indian raids, but because the Arizona Copper Company was in financial distress brought on by several problems. Of great importance was the cost, which was much more than expected, and there was a cave-in at the Longfellow Mine. Close to bankruptcy, the company rescued itself temporarily by issuing new bonds and reorganizing under a new board of directors.

Work on the railroad later resumed and, in 1883, James Colquhoun, who later became president of the Arizona Copper Company, rode on that railroad to the point where the rails then ended, at Guthrie. During the financially trying years of 1884 to 1888, Lawrence Russell was the general

---

[3] William Ryder Ridgway, *Mount Graham Profiles, Ryder Ridgway Collection,* Graham County Historical Society, 1988, 244.

manager. He struggled to regain the company's footing after its near-collapse.

The Arizona and New Mexico Railway, running from Lordsburg and all the way to Clifton, was eventually finished as a narrow gauge line (36 inches between rails) and was later widened to standard gauge (57 inches between rails).[4] Regular passenger service was an important feature.

At that time, Phelps Dodge owned and operated the Detroit Copper Mining Company. Detroit Copper had contracted with Arizona Copper to use the A & NM Railway.

In 1908, James Douglas, president of Phelps Dodge, approached James Colquhoun, president of Arizona Copper, with the proposal of extending the A & NM Railway from Lordsburg to Hachita, New Mexico to connect with the El Paso and Southwestern Railroad which was owned by Phelps Dodge. If the A & NM were so extended, Phelps Dodge could avoid the Southern Pacific's freight charges.

Colquhoun agreed and the A & NM was extended from Lordsburg to Hachita.[5] Regular passenger service then was available from Clifton to Hachita.

In the 1920s the A & NM was sold to the Southern Pacific Railroad and the portion of the line between Lordsburg and Hachita was abandoned. Eventually, the EP & SW railroad was also abandoned.[6]

The Southern Pacific continued passenger service from Lordsburg to Clifton up to the 1960s. The last car on the train was a half-baggage, half-passenger combination which trailed along behind the freight train in and out of Clifton each day. My first trip to Morenci ended on that car from Lordsburg to Clifton in 1951. I was the only passenger and the first one in a long time.

The Morenci Southern Railroad

In 1901, the Morenci Southern Railroad was born. It was built by the Detroit Copper Mining Company (owned and operated by Phelps Dodge since 1897) and covered 18 miles from Guthrie to Morenci. This was a

---

[4] Carmichael and Kiddie, 832.

[5] Colquhoun, 33.

[6] *Phelps Dodge, A Copper Centennial, 1881-1981, PAY DIRT, Summer 1981,* William C. Epler, Editor, 134-138.

narrow-gauge railroad with regular passenger as well as freight service. The first run was held in January 1901.[7]

Morenci Southern was famous for its spectacular loops each of which looked like a giant roller coaster. The train circled carefully around each loop to gain the necessary elevation into Morenci. The railroad had seven steel bridges, numerous timber trestles, three tunnels and five complete loops.[8]

Remember the old "Black Bridge" that crossed the San Francisco River below the corrals? It served the Morenci Southern Railroad and was a fascinating point of interest long after the railroad was abandoned. Unfortunately, the bridge was also a safety hazard, luring people to try to cross it and to swing on it at the end of a rope. To avoid any accidents, it was destroyed during the early 1980s.

The south end of the railroad terminated at Guthrie where it joined with the A & NM Railway at the Gila River. The junction was located on the west side of the river in the vicinity of the present Southern Pacific Railroad bridge at Guthrie, near the location of Dick and Frances Powers' ranch.

Because of the loops, the railroad was named "The Corkscrew Route of America."[9] It was narrow gauge (36 inches between rails). Eventually, the condition of the wood trestles deteriorated, causing concern for the safety of the trains and they were replaced with switchbacks.

The Morenci Southern Railroad was abandoned shortly after Phelps Dodge acquired control of the entire district in 1921.

The Shannon Railroad

In the early 1900s, the Shannon Copper Company built a railroad from the Shannon mine, located north of the town of Metcalf, to their mill and smelter located on Shannon Hill in Clifton. They had been using the Coronado Railroad owned by the Arizona Copper Company up to that time.[10]

The narrow gauge (36 inches between rails) Shannon Railroad started from Clifton, then followed Chase Creek Canyon, paralleling the Coronado Railroad. Overlooking Chase Creek, the Shannon was carved out of the cliffs

---

[7] Frederick Remington Barr, *The Copper Era, Victory Resource Number,* April 21, 1943, 10.
[8] Roberta Watt Troxell, *History of Morenci, Arizona,* Master's thesis, University of Arizona, 1956, 87.
[9] Troxell, 89.
[10] Carmichael and Kiddie, 826.

above the Coronado Railroad. On up Chase Creek Canyon, it passed through Metcalf and ended at the foot of the Shannon incline. One impressive trestle was built in the middle of the town of Metcalf where the railroad crossed Santa Rosa Canyon.

There was no scheduled passenger service on the Shannon Railroad but passenger service was provided for special events. Since the railroad passed by the Catholic cemetery, trainloads of mourners would be transported there and back to Clifton for funerals.

Mine Haulage by Electric Locomotive

In 1905, the Arizona Copper Company replaced mules with electric locomotives on the surface and important levels of the Humboldt Mine. Beginning in 1906, small electric locomotives hauled ore from the Humboldt Mine, largest underground mine in the District, to the No. 6 concentrator. Ore cars ranging in size from one ton to ten ton capacity were used on baby gauge (20 inches between rails) track. A company engineer did what the companies that built locomotives could not do. He designed a system where three electric locomotives could be hooked together, operated by one engineer instead of three. The need for three locomotives came about with the desire to increase the tonnage being hauled. The three locomotives pulled a 300-ton ore train.

The Matilda shaft of the Coronado Mine was connected to the rest of the District only by the short railroad that ran along the top of Coronado Mountain to the spectacular Coronado incline. In 1912, a mammoth undertaking connected the Coronado incline to the Clay shaft of the Humboldt Mine by extending the electric haulage system through 8,000 feet of tunnels and 6,600 feet of rail bed along the mountain sides.[10] Tony Enrico, who was born and raised in old Morenci and later became Phelps Dodge Morenci Branch chief design engineer, often spoke of walking through one of those tunnels on his way from Morenci to the town of Metcalf.

---

[10] Carmichael and Kiddie, 835.

Narrow gauge train on a loop of the Morenci Southern Railway in the early 1900s. The town of Old Morenci appears in the upper left corner. *Courtesy Roberta Troxell.*

Well dressed travelers pose for the camera on a baby gauge passenger train at the Clifton Railroad station. Picture probably taken in the 1880s. *Courtesy Phelps Dodge Corporation.*

Train crewmen pose beside their narrow gauge (36 inches between rails) locomotive No. 20 on the Coronado Railroad in Clifton. *Courtesy Phelps Dodge Corporation.*

Locomotive No. 1 at rest in the Clifton railroad yard.
Courtesy Roberta Troxell.

A baby gauge locomotive pulls ore cars near Metcalf in the late 1800s.
Courtesy Phelps Dodge Corporation.

A tunnel on the Coronado Railroad shortly after 1900. This area has long been mined out. The railroad bed's elevation would now be several hundred feet in the air. *Courtesy Phelps Dodge Corporation.*

Narrow gauge saddle tank locomotive used in the Clifton-Morenci area in the early 1900s. *Courtesy Phelps Dodge Corporation.*

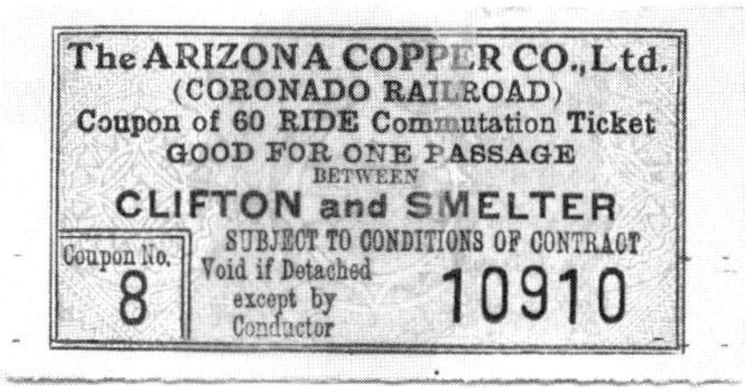

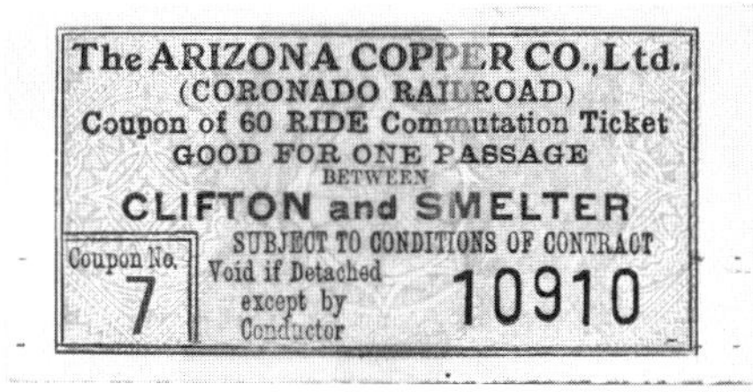

Train tickets such as these were issued to passengers by the Arizona Copper Company. *Tickets Courtesy Al Fernandez.*

Old narrow gauge passenger coach on the Coronado Railroad at rest near the Clifton station. *Courtesy Phelps Dodge Corporation*

A unique camera shot of a narrow gauge slag train on the Coronado Railroad beside a standard gauge box car on the Arizona and New Mexico Railway at the Clifton rail yard in about 1910. *Courtesy Phelps Dodge Corporation.*

16

The great little baby gauge locomotives of Clifton-Morenci.... TLC

In 1878 and the early part of 1879, Henry Lesinsky, principal owner of Longfellow Copper Company, constructed a railroad between the Longfellow mine in Morenci and his smelter in Clifton.[1] This was not the first railroad in Arizona, as some have said. The first Arizona railroad had appeared about two years earlier, building eastward from Fort Yuma. But the railroad Lesinsky and associates built and called "Coronado" was the first mine railroad in Arizona. Ore cars on this railroad were pulled upgrade by mules when empty. Loaded, they coasted down the canyon to the smelter, with the mules and the brakeman riding in the last car.

The great improvement of steam locomotion occurred in 1880 when Lesinsky purchased a baby-gauge locomotive from the H. K. Porter Locomotive Works in Pittsburgh, Pennsylvania. Riding on rails that were only 20 inches apart as opposed to a standard track's 56 and one-half inches, a baby gauge was tiny.

That locomotive was small, but it did weigh about 9,000 pounds, which was considerable considering the transportation of the time. In order to ship it, it had to be disassembled and boxed so that, according to *The Railroad Gazette* of June 20, 1880, it could be "hauled by wagon several hundred miles from the railroad terminus." Just getting that first mine locomotive to Clifton had to have been quite an accomplishment. Part of the shipment consisted of a box weighing 2,000 pounds with dimensions of 8 x 4 x 3ft. This contained the frames, cylinders and other items. Shipped in April 1880, the locomotive left New York City by steamer and went around the Horn to San Francisco. From there it was by rail to Fort Yuma and then by ox-team to Clifton.[2]

---

[1] Norman Carmichael and John Kiddie, *Transactions, American Institute of Mining and Metallurgical Engineers, Vol.LXX, 1924,* "Development of Mine Transportation in Clifton-Morenci District," 829.

[2] Carmichael and Kiddie, 831.

At the H.K. Porter factory in Pittsburgh, they painted the name "Coranada" on it in an attempt at spelling "Coronado." Upon its arrival in Clifton, its engineer, Henry (Dad) Arbuckle, helped assemble it; and its name was changed to "Little Emma," after one of his daughters.[3]

Dad Arbuckle, rightly proud of his steam engine, described for the *Clifton Clarion* of February 18, 1885, the hair-raising adventure of an attack by a "band of fifteen or twenty Apaches" on the first day he operated Little Emma or "El Vapor" as the local population called it. Although the section gang and Arbuckle had what we'd call a "close shave," the Apaches failed to hurt them or the little engine; and it was, in fact, to protect against such attacks against the more helpless teamsters with ore wagons as well as the wagon-haulage cost of $10 per ton of ore that Henry Lesinsky had purchased the baby gauge.[4]

Lesinsky would later purchase a second baby gauge locomotive in that same year of 1880 at a cost of $2,600. Records indicate this one was to be shipped by steamer from New York to San Francisco. In 1882, Lesinsky ordered a third baby gauge for $5,950.[5] These marvelous little machines, which represented the cutting edge of mining technology in the late 19th and early 20th centuries, would fit in the shovel dippers of today's Morenci Mine.

Also in 1882, Henry Lesinsky and his associates sold the Longfellow Copper Company to the Arizona Copper Company, a firm headquartered in Edinburgh, Scotland. Arizona Copper expanded the former Lesinsky fleet with the purchase additional baby-gauge locomotives. Detroit Copper Mining Company, which also operated in the Clifton-Morenci district at that time, purchased baby-gauges for their operations.[6]

One of the baby-gauge locomotives, Arizona Copper's No. 8, an 1897 model, was restored by Tommy Sidebotham in 1937 and has been on display in Clifton since that time. Another is in a museum in Phoenix. With the exception of three others, Arizona Copper's Nos. 2, 5, and 7 locomotives, all the rest disappeared, probably in the end to World War II scrap metal drives.

For 67 years, the only remaining baby gauges at the mine site were Nos. 2, 5, and 7. Fortunately, for the sake of history, somewhere in the early part of the century the three had been parked atop rugged Coronado Mountain at

---

[3] Carmichael and Kiddie, 831; and William Ryder Ridgeway, "Whistling down the Canyon," *Arizonan Highways,* July 1961.
[4] Carmichael and Kiddie, 829; and Ridgeway.
[5] Ridgeway.
[6] For an authoritative discussion of the Clifton-Morenci locomotives, see David F. Myrick, *Railroads of Arizona, vol. III,* Trans-Anglo Books, Glendale, California, 1987.

the northwest side of the Morenci open pit. There was no way of bringing them down, so there they stayed. Considering how inaccessible was the site, we wondered how they came to be placed there. Our best guess: they had probably been pulled up there in sections on the Coronado incline and then reassembled.

Phelps Dodge purchased the Arizona Copper holdings in 1921 and operated the old ACC underground mine on Coronado Mountain until 1923, using those three baby-gauge locomotives on a railroad of about a mile's length that had been constructed on the mountain to haul ore from the Matilda shaft of the Coronado Mine to the top of the Coronado incline.

When mining stopped in 1923, the three baby gauges were neatly parked on small sections of track near the Matilda shaft where they gathered rust and lost to souvenir seekers parts that could be carried away. Gone were the whistles and the bells and, most importantly, from a historical perspective, the brass builder's plates and the road numbers (the number on the very front of the engine). Thus there was no way to immediately identify the locomotives from the remaining evidence.

Over the years, word of the three historic engines got out and many plans were suggested by sources outside Phelps Dodge for their removal from that mountain hideaway. One of these involved mounting the engines on skids and sliding them down the mountain on the steep bed of the old Coronado incline from which the rails had been removed in the 1930s. But no one knew how the speed of that ride could be controlled.

In another proposal, the engines would be airlifted off the mountain with a helicopter. The group making that suggestion induced the pilots of a helicopter to fly to the mountain top in a reconnoitering mission. After finding a safe place to set down, which wasn't easy in that terrain, the pilots wisely concluded that the density-altitude penalty of that high elevation, 5,700 feet, would be too severe for the lift. Like Little Emma, they were baby gauges, but were bigger than Little Emma and were much heavier, one reason being the immense weights that had been secured to the front of the engines to improve traction. Number 5, as an example, weighed 28,000 pounds. Until March of 1990, none of the removal plans would prove to be feasible.

Then, early in the new year of 1990, Fred Menzer's PDMI geology department had a road constructed up Coronado Mountain to accommodate drilling equipment that would be used to define the Coronado ore body for future mining. This road, though perfectly adequate for hauling drilling equipment, had not been designed for bringing locomotives down a mountain. Still, we wondered if it could be used to do just that. I decided to take a close look at it, and Bob North of the geology department took me up the mountain. The road was definitely narrow and had many interesting switchbacks which, as we approached the top, provided great views of that winding trail far below that we had just left. Bringing those engines down on that road did seem possible, but it would not be any piece of cake.

When I returned I brought up the possibility with Jeffrey Clevenger, PDMI president. As was typical of all my dealings with him, Jeff lost no time in making his decision. His directive to me was not complicated: "Bring them down off the mountain," he said.

I was particularly pleased with that decision. The three tiny engines had long been of special interest to me. I had climbed the rugged and steep mountain three times to see them with my daughters and sons-in-law. You couldn't, in my opinion, visit those beautifully crafted reminders of a long-past mining era too many times. And you couldn't avoid being moved by the sight of them — resting on short lengths of tracks in quiet dignity on the solitary mountain in exactly the same place where their fires were allowed to die way back in 1923.

This was a tiring climb, a decidedly steep ascent in which you would elevate yourself a full 600 feet. But not as severe as the 1,200 foot rise that had existed when the incline was in use. Over the many years since the incline's abandonment, barren material from the mine had been deposited at its base and Highway 666 (later 191) had been built across it, reducing the vertical rise by about a half.

On my last hiking trip up that mountain, we went up an adjacent canyon which was not as steep as the Coronado incline. Just as I had taken the lead, (my son-in-law Al Swartz was with me) I all but stepped on a rattler who thought he was the owner of that particular shady spot on the uphill trail. He was neatly coiled with the first foot of his long body extended upward when I saw him, my boot only about a foot above his head. We slipped back, carefully climbed up around Mr. Rattler and, except for what seemed like a vastly accelerated heart beat, no harm was done. The rattler slipped away beneath a boulder as we went around him. (On our way back down, we were looking for him when we neared that location and found him occupying the exact spot on the trail where we had found him on the ascent. Once again, we gave way.)

At last we reached the baby gauges' elevation. The building that had once covered them, and had provided the mechanics who worked on them some shelter, had probably disappeared more than half a century earlier. The additional exposure to the elements had increased their deterioration — but the tiny engines also made you think they had done their best to resist the ravages of the weather. They were still impressive reminders of that bygone age of steam. Though there was hardly a sign of it, there had once been other buildings there, even a few houses. The metal headboard of a bed, an ancient cooking stove, a broken piece of dinner plate bore evidence of habitation.

In that absolute quiet of the deserted mountain you felt as if you had stepped into some science-fiction time machine and had emerged in the early nineteen twenties. And you couldn't help but think Coronado Mountain seemed as much the proper habitat of those tiny locomotives as the adjacent canyon had been the rattler's. We knew it was ridiculous, but we thought it would be nice if the engines could just stay there for all time.

Retrieval of the locomotives was to be undertaken by Jack Cooper's mechanical department crew. I would not have said anything except to agree that it clearly was an impossibility and would add that I didn't really think it could be done but merely wanted their concurrence — if that had been their conclusion. But Brad McBiles, now retired and then mechanical-electrical division superintendent, and Jack Cooper and his crew assured me that it could be accomplished.

And they did it. They did it by loading the engines, one at a time, onto a flatbed truck that they drove — very slowly — down the mountain. In places where the road was exceptionally steep, the flatbed was sort of lowered down the mountain by means of a cable attached to a trailing dozer. Tying the truck to the dozer was important. If that had not been done, the truck with its important cargo might have become a runaway and gone crashing down the precipitous slope. In a short time, it was all performed according to plan, and I received permission from Jeff Clevenger to have one of the locomotives restored for display in the Morenci Plaza.

After considering the advantages and disadvantages of restoring the other two, I decided to restore "Number 5." It was the only one of the three for which we knew what the bell, whistle and road number had looked like, items that had disappeared from all three engines a long time ago.

In the Phelps Dodge Morenci photography department there was a glass plate negative of one of Arizona Copper's baby-gauge locomotives. That picture, taken about 100 years earlier when the locomotive was operating in the long-gone town of Metcalf, an area now part of the open-pit mine operations, clearly showed the engine's road number — the number 5. This was the only old photograph that could be used to identify any of the three locomotives. By comparing the rivets and bolts shown in the photo with those on the actual engines, we were able to identify one of the three that had been resting 67 years with boilers cold on Coronado Mountain as ACC Number 5.

But we still did not know what was on the builder's plate — that shield-shaped object made of brass that had been attached to the side of the engine and had disappeared along with the bell, whistle and road number. We asked Ken Malloque of the photography department to enlarge that part of the photo. The engine's serial number and its year of manufacture — 1888 — miraculously emerged from the gray background. But only a few letters in the name of the superintendent of the Porter Locomotive Works, a name that had been on the builder's plate, were distinguishable in the enlargement.

After a number of back and forth calls with the Carnegie Library in Pittsburgh, I received a page from the publication *Industries of Pittsburgh,* which was written in 1879. A brief description of the Porter Locomotive Works appeared in that publication along with the superintendent's name — "H. N. Sprague." The number of letters in that name was the same as had been on the builder's plate and the few that were distinguishable in the enlarged photograph matched perfectly as to their place in the name "H. N. Sprague." With that information, and a builder's plate "go-by" from the Greenlee County Historical Society museum that was on loan from Bill Cochran, a perfect reproduction of Number 5's builder's plate could be made.

After a couple of months of what must have been a labor of love, Neil Hise of CEMCO Manufacturing in Albuquerque made Number 5 look as it probably did when it was rolled out of the H. K. Porter factory in Pittsburgh more that a century before. He completed the restoration in accordance with what he called "our blueprint" — the photograph taken about 100 years ago — and the locomotive now had a whistle, a shiny brass bell, and a historically accurate brass builder's plate. Number 5 was placed on display in the Morenci Plaza on January 18, 1991 where visitors now enjoy a close look at one of those tiny workhorses from the age of steam.

The other two baby gauge locomotives retrieved from Coronado Mountain on that March day in 1990 were Number 2, an 1882 model, and Number 7, an 1896 model. Number 2 was also restored by Neil Hise and was placed on display at the new Arizona Mining and Minerals Museum in Phoenix. On September 15, 1993, it was dedicated in ceremonies also honoring the retirement of Leroy Kissinger as Director of the Arizona Department of Mines and Mineral Resources.[7] Number 7 has been in Safford a number of years, dismantled and awaiting restoration. Phelps Dodge donated Number 7 with the condition that it be restored and placed on display at the Museum of Discovery in Safford.

---

[7] Former Arizona Governor Rose Mofford and long-time State Representative Polly Rosenbaum, both with copper mining in their family backgrounds, were among the speakers. Mason Coggin , who was from the mining town of Bisbee, succeeded Leroy Kissinger as Director of Mines and Mineral Resources.

An artist's rendition of the first locomotive in Clifton-Morenci. It was meant to be named "Coronado" but the spelling of the name of that explorer of the West was probably unfamiliar to the folks in the East where the locomotive was built (Pittsburgh). When put in service it was named "Little Emma" after the engineer's daughter and called "El Vapor" by the townsfolk. *Courtesy Tom Rees from the Dave Rees collection.*

One of the early baby gauge locomotives was named after Henry "Dad" Arbuckle who had been the engineer on "Little Emma," the first mine locomotive in Clifton-Morenci and in Arizona. *Courtesy Tom Rees.*

Picnickers pose for the camera beside a baby gauge locomotive of the 1890s era at one of the Coronado Railroad tunnels between Metcalf and Clifton. *Courtesy Tom Rees.*

Baby gauge locomotive No. 5 at Metcalf in about 1890. When Neil Hise restored this locomotive at his Albuquerque shop in 1990, he used this photo as his "blueprint." *Courtesy Phelps Dodge Corporation.*

A baby gauge locomotive of the 1890s era and its crew rest below the famed Coronado incline. *Courtesy Phelps Dodge Corporation*

A baby gauge locomotive moves ore along a railroad of about a mile's length that leads from the Matilda shaft of the Coronado Mine to the top of the Coronado incline. The shape of the cab roof shows this tiny engine to be either No. 2 or No. 5.

Courtesy Phelps Dodge Corporation.

The only way to get a good camera shot of all three baby gauge locomotives while they were marooned on Coronado Mountain was to climb up onto the bluff above the old railroad bed. In this shot, No.2, an 1882 model, is in the front of the picture, No.5, an 1888 model, is on the left and No.7, an 1896 model, is at the right. *Photo by Ted Cogut, 1986.*

In March 1990, Phelps Dodge Morenci mechanical/electrical personnel examine the three baby gauge locomotives to determine the feasibility of moving them down off Coronado Mountain. Shown here are No. 5 and No. 7 (flat roof). *Photo by Jack Cooper.*

Precious cargo, No. 5 baby gauge locomotive, an 1888 model, is carefully swung onto a flatbed. *Photo by Ted Cogut, 1990.*

Neil Hise, walking, prepares to secure baby gauge locomotive No. 5 to the flatbed for its trip to Albuquerque and a face lifting at his CEMCO plant. *Photo by Ted Cogut, 1990.*

Baby gauge locomotive No. 5, fresh from its restoration in Albuquerque, being unloaded on January 18, 1991 in the Morenci Plaza and permanently displayed there. *Photo by Ken Malloque.*

Local detective work assisted by information from the Carnegie Library in Pittsburgh enabled Neil Hise of Albuquerque to produce this replica of the builder's plate for baby gauge locomotive No.5. *Photo by Ken Malloque.*

The Fourth of July, 1991. No. 5 was asked to go to the parade in Duncan, Arizona. She went there and has since participated in a number of other events where a good looking locomotive is in demand. *Photo by Ken Malloque..*

Baby gauge locomotive No. 7, an 1896 model, pictured after it was stripped down by volunteers working at its restoration in Safford, Arizona. One of three baby gauge locomotives removed from Coronado Mountain in 1990, No. 7 was donated to the Museum of Discovery in Safford, Arizona with the provision that it be restored and put on display there. In the event that the restoration is not to be accomplished, the locomotive is to revert back to Phelps Dodge ownership. *Photo by Pat Mahoney who had worked on the restoration until a shortage of funds brought a stop to the effort.*

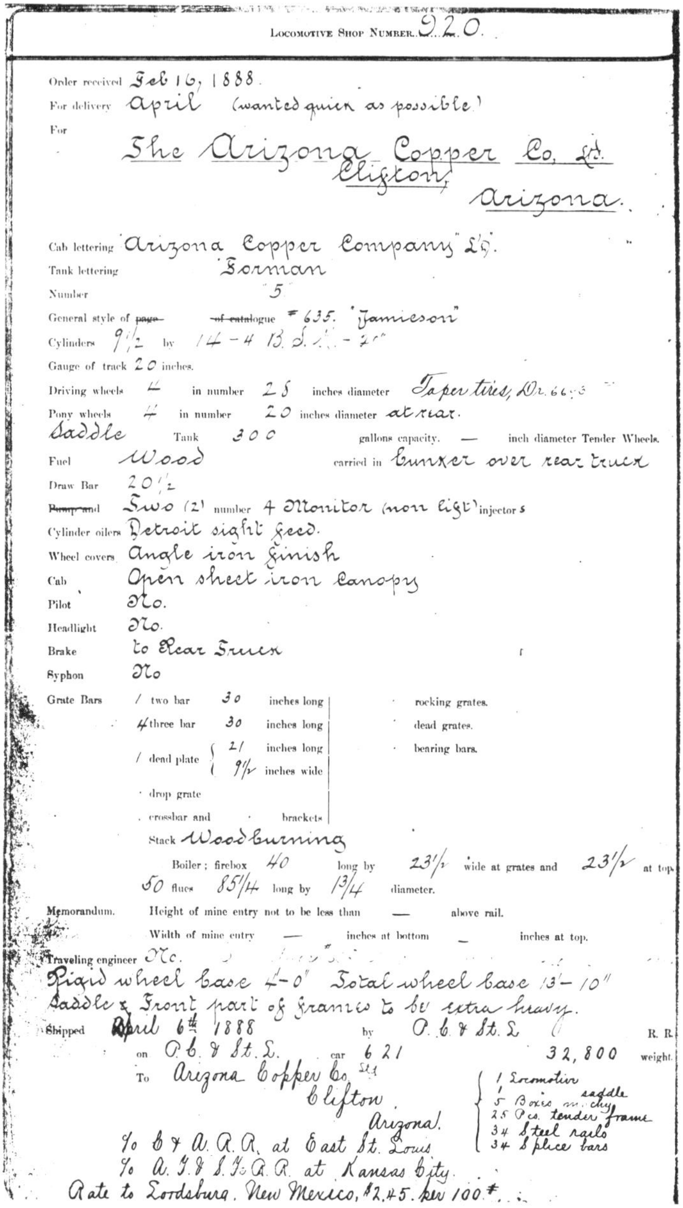

Locomotive Shop Number 920.

Order received Feb 16, 1888.
For delivery April (wanted quick as possible)
For The Arizona Copper Co. Ld. Clifton, Arizona.

Cab lettering "Arizona Copper Company" Ld.
Tank lettering "Forman"
Number "5"
General style of ~~page~~ ~~of catalogue~~ # 635. "Jamieson"
Cylinders 9½ by 14 - 4 B. S. - 2"
Gauge of track 20 inches.
Driving wheels 4 in number 28 inches diameter Taper tires, Dr. 66
Pony wheels 4 in number 20 inches diameter at rear.
Saddle Tank 300 gallons capacity. — inch diameter Tender Wheels.
Fuel Wood carried in Bunker over rear truck
Draw Bar 20½
~~Pump and~~ Two (2) number 4 Monitor (non lift) injectors
Cylinder oilers Detroit sight feed.
Wheel covers Angle iron finish
Cab Open sheet iron canopy
Pilot No.
Headlight No.
Brake to Rear Truck
Syphon No
Grate Bars 1 two bar 30 inches long · rocking grates.
4 three bar 30 inches long · dead grates.
1 dead plate 21 inches long, 9½ inches wide · bearing bars.
· drop grate
. crossbar and · brackets
Stack Woodburning
Boiler; firebox 40 long by 23½ wide at grates and 23½ at top
50 flues 85¼ long by 1¾ diameter.
Memorandum. Height of mine entry not to be less than — above rail.
Width of mine entry — inches at bottom — inches at top.
Traveling engineer No.
Rigid wheel base 4'-0" Total wheel base 13'-10"
Saddle & Front part of frames to be extra heavy.
Shipped April 6th 1888 by O. C. & St. L. R. R.
on O. C. & St. L. car 621 32,800 weight.
To Arizona Copper Co. Ltd
Clifton
Arizona.
1 Locomotive
5 Boxes machy saddle
25 Pcs. tender frame
34 Steel rails
34 Splice bars
% C & A. R. R. at East St. Louis
% A. T. & S. F. R. R. at Kansas City.
Rate to Lordsburg, New Mexico, $2.45. per 100#.

The factory order for baby gauge locomotive No. 5.

Courtesy Vernon and Ellen Glover, Albuquerque.

17

Pete Gomez began working at Phelps Dodge Morenci in 1922 and stayed 44 years.... WCC

Pete Gomez is an amazingly active, intelligent man who will be 94 years old this year (1997). I spent about four hours with him as he recounted events in the early years of Morenci. I really only recorded a fraction of his fascinating story. He can recall in detail events that happened many years before most of us were born. This chapter provides some of the highlights of our interview.[1]

Pete was born in Jalisco, Mexico, in 1903. In 1905, his family brought him to the U.S. through Juarez and El Paso to a town called Smelter, which is now a part of El Paso. There were no immigration laws then, and no stopping at the border. Juarez and El Paso were small towns. Pete's father went to work in a smelter in El Paso for 10 cents an hour. He made $ 1.20 per day, working 10-hour days.

The railroads were expanding all over the West. They would provide free train fare to those seeking work. Many of the people from Mexico would work on farms. Some went as far as California and Colorado. The railroad, Pete said, would "dump people in the middle of nowhere." Pete and his family rode on the train to Dawson, in northern New Mexico, where Phelps Dodge worked a coal mine.

At Dawson, it was impressively cold in winter, but their house was very comfortable, built of railroad ties. For fuel, the workers were allowed to pick chunks of coal from the railroad cars and burn old railroad ties. Pete's dad worked at Dawson for six months (six-month contracts were pretty much standard procedure), then back to work in the smelter in El Paso.

Pete's family learned about Morenci through his grandmother and uncle who lived there. So they moved to Morenci. That was in 1910, and Pete lived in the Morenci/Clifton area until 1996 — 86 years! When they came to Morenci, there was no electricity for the homes. In the frame houses

---

[1] The Pete Gomez story first appeared in the February 1997 issue of *Morenci Copper Review*.

of that time, "you could see out between the cracks. Imagine what it was like in winter!"

Some of the girls and boys in Pete's second-grade class were 16 and 17 years old when Pete started school in Morenci. That was in the Longfellow School, which was located next to the Morenci Hotel. Besides regular school, Pete went to Spanish school in the summer. Professors from Mexico would teach Mexican history and other subjects.

Pete talked about his favorite teacher, Mrs. Barr. She was the wife of F. Remington Barr, who later became Morenci Water and Electric superintendent and a renown historian of the Morenci district. He said Mrs. Barr was a tiny woman, but one stern look from her was like an electric shock! When you went into her classroom, you could hear a fly and the tick-tock of the clock. She had discipline! Pete was one of her pets, partly because he was an avid reader and because he was good at translating from Spanish to English.

Most people dropped out of school because of hardship. Morenci was lucky to graduate even one person in a year. That person could pick just about any job he wanted. Most of the boys then went to work in the mines.

After finishing sixth grade, Pete had to quit school to help support his family. Copper prices were badly depressed in 1914, and work for Pete's dad was hard to find. Copper was selling for 16 cents per pound or less.

At that time, there were nine members of the family. His dad, who was employed by both the Arizona Copper and Detroit Copper companies, was only working five hours per day, five days per week. The pay for a whole day of work was $1.60 for surface work and $2.25 for underground work. Sometimes, when things got bad, you only made five dollars a week for a five-day week. Pete says "there was no welfare in those days. You had to work to make a living!"

In 1920, when Pete was 17, he and his father did something that was considered no big deal in those days but almost impossible today. They walked across the mountains to Safford, then took a train to Miami in search of his dad's brother.

They couldn't find him there, so they then walked across more mountains to Ray. There they stayed two or three days, then headed for Superior, again on foot, and found the object of their search playing pool in the Sunset Pool Hall.

What a reunion took place! Pete's dad cried "There's my brother!" The uncle was so surprised he threw his cue stick across the table.

Finding work, they stayed in Superior three months. Then they headed home, hiking across the mountains between Superior and Miami where prisoners were building the first road that would connect the two towns. At Miami, they boarded the train to Safford and then trekked on foot back to Morenci.

One year later, Pete took another hike. He walked from Morenci and up the Frisco River to the town of Mogollon, New Mexico where he stayed for three months. It took him two days to cover the distance. We can appreciate what he did when we realize it's about 80 miles by road from Morenci to Mogollon! Nowadays, we complain about how long it takes us to drive to Tucson or Phoenix.[2]

Mules then were the main transportation in Morenci. They would pull five ¾-ton cars full of ore out of the underground mine. Pete describes them as big, beautiful animals. One man would train a mule, and the mule would obey his commands.

The mule would come out of the mine hauling the cars at a run and would then stand to one side of the track as the miner would dump the cars. Then the mule would come back to the cars on his own, ready to be hooked up and go back to work. Many of us remember the syrup cans shaped like log cabins. One of these cans with a candle in it would be hung around the neck of the mule.

Sometimes a mule would be in an underground mine for many months. When he would finally be brought out, it would be at night. Then he would be put in a dark room with only candle light. Slowly, each day, the door to the room would be opened a little more until the mule would become used to daylight. He would then be turned loose on the tailings area to live out the rest of his days, having served his time underground.

Pete went to work for Phelps Dodge in September 1922 as a boilermaker helper. He was ambitious and wanted to learn as much as he could. At that time, the starting pay for boilermaker helper was $2.90 per day. The top of the line was journeyman special at $5.50 per day.

Miners from many different countries had been working at Morenci. There were quite a few people from England and Scotland still there who had

---

[2] For the reader unfamiliar with the area, it is about a 3-hour drive to Tucson and 4 to 4 1/2 hours to Phoenix.

worked for the Arizona Copper Company which was based in Edinburgh, Scotland. One of these with whom Pete worked was the master mechanic, Mr. MacDugal. The general foreman, Dominic Maffeo, was of Italian descent. Maffeo, a very good boilermaker from Bisbee, gave Pete the opportunity to learn the trade.

In those days, many boilermakers didn't stay in one place very long. They roamed all over the country, riding the rails on freight trains. They'd stop in Morenci for a while to work, then move on. Since they weren't dependable, Maffeo needed to train someone who was and Pete filled the bill. Pete's first specialty was steam locomotive repair. The locomotives were used on the Industrial Railroad between Clifton and Morenci, which at that time was narrow gauge. Pete worked his way up to journeyman, making $4.80 per day.

It was about the time the Morenci Southern and Coronado railroads were being phased out, and the Arizona and New Mexico Railroad was soon to become the Southern Pacific.

In an early edition of the *Morenci Copper Review,*[3] we ran an article about Pete and some of the safety awards he won during the 1920s. A safety program was started in 1925. Every department was in competition for fewest accidents, and a banquet would be held for the winning department. Pete's department, the machine shop, won three annual awards.

Everybody went home at noon, got cleaned and dressed up, and came back for the food and awards at 4 p.m. A long table was set up in the shop, graced with a huge pot of beans and ham. There was a drawing, a movie was shown and everyone dined. Awards were presented. Everyone got something during the drawing. Some of Pete's awards were a belt buckle, a knife and a watch fob. The watch fob was presented in 1928.

We think of solution extraction-electrowinning as a new development in the business of making copper. Not so, says Pete.

In 1927, Phelps Dodge had a pilot plant at the No. 6 concentrator. There were two concrete tanks filled with low-grade, crushed copper-bearing rock. Lead pipes were used to run water from tanks inside the concentrator, which were also made of lead, to the outdoor tanks containing the low-grade ore and then back in a continuous circulation. In the tanks inside the concentrator, plates hanging from a beam were submerged in water. Electric wires were connected to the plates and sulfuric acid was added to the water.

[3] The November 1987 issue of *Morenci Copper Review.* Close-up photos of the front and back of Pete's safety award buckle were included with the article.

Copper adhered to the plates, producing a thin, beautiful red 100% copper sheet. This plant was shut down when the testing was done.

In 1929, Pete and Marina Lopez were married. Marina and her mother were both born in Metcalf. Her mother in 1889. Marina's grandfather came to Metcalf in 1882, when everyone lived in tents. In those days, the miners worked with a rifle handy because of Indian raids.

A group called Club Verde was started in Metcalf in 1914. Marina's father was one of the organizers. They would meet every year for a celebration. This tradition continues even to this day with descendants of the old towns of Metcalf and Morenci coming back from California and elsewhere to celebrate in Clifton. Up to 1983, busloads would come in from California. I remember them meeting at the Oasis on Clifton's east side.

In 1932, during the Great Depression that had begun in 1929, all mining came to a close in Morenci. During the five-year shutdown, people worked for the government's C.C.C. (Civilian Conservation Corps) camp. The pay was $44 per month.

Afterward, Pete went back to work in the machine shop where his specialty became shovel repair. He had learned to read prints during his work assignments in the No. 6 concentrator expansions of the late 1920s and became a boilermaker layout man who actually drafted out the plan of work on boilerplate. He built dippers and other shovel parts.

Pete retired from Phelps Dodge as a boilermaker gang boss after 44 years of dedicated service. He has seen much of the history of Morenci, and he feels that history books don't give credit to the brave men and women who pioneered this area. There were no roads, no railroads, no trails. What tremendous sacrifices they made.

Pete and Marina have been married for 68 years. They now live in Safford with their daughter and son-in-law, Valia and Simon Peru. They also have four grandchildren and six great grandchildren.

Pete believes strongly in the importance of a good education. He is outspoken about the value of hard work and willingness to learn. Pete's successful life and work record bear strong testimony to his beliefs.

Ore cars loaded with ore from the Longfellow Mine are about to be lowered down the Longfellow incline.
Courtesy Phelps Dodge Corporation.

When Pete Gomez arrived in Morenci, burros were an important means of moving loads of all sorts and were well taken care of by their owners. *Courtesy Roberta Troxell.*

Pete Gomez in 1997. *Photo by Bill Conger.*

18

They weren't all big companies.... WCC

Besides the three major copper companies (Arizona, Detroit and Shannon) in the early days of the District, numerous small companies also mined copper. Some of these were the New England, Clifton Consolidated, Stevens, King, Queen and Standard. They mined copper with varying degrees of success. Many years after the copper veins of the King and Standard ore bodies were mined out and abandoned, mining started again; and those areas are now part of the huge Morenci-Metcalf open pit complex. The old Queen Mine later became the Queen Hill section of the open pit Morenci Mine.

Granville was another early camp. It was mined for silver in the 1890s. This became another example of resuscitation of an old mining area, but this time mining did not resume. Located on the west side of the Coronado Trail which ran along the west side of upper Chase Creek, it was six miles north of the Morenci open pit. Granville became the home of Mr. and Mrs. Commeau, Roy and Francelle. Many of us who lived in Morenci or Clifton knew the place as "Cherry Lodge."[1] It was one of those spots at the side of the road that beckoned one to linger awhile. At times you'd have to slow down or stop as you went by to make way for chickens that had wandered from the yard. Scouts and other youth groups often went there on outings. The Commeaus rented out cabins and provided dinners by reservation. You could call from Morenci or Clifton requesting a chicken dinner and when you arrived at the lodge to enjoy it you knew you were being served the freshest chicken in the West. Your dinner had been out running in the yard at the time of your call.

There was a tunnel (adit) that opened directly onto the shoulder of old U.S. Highway 666, a road had been carved out of the cliffs alongside Chase Creek and, at that time, took the traveler along the east side of the Morenci open pit. The tunnel was a focal point of interest to everyone passing by it. We called it the "Christie Tunnel" of the Western Copper property.

---

[1] Francelle Commeau was deputy assessor at the Greenlee County courthouse. After her husband Roy passed away, she moved to be near relatives in Charlotte, Michigan. In a conversation with Ted Cogut in 1991, she said she deeply missed the Clifton-Morenci area. Francelle Commeau passed away in the mid-1990s. On one deep-snow day in the winter of 1974-75, Francelle became a heroine to a member of the Phelps Dodge environmental department whom she rescued after he became snowbound while servicing the Cadillac Point air monitor at the north edge of the open pit. The reader will no doubt guess who the lucky person was. The story appeared in the "Cloud Chasing" column of the November 1991 issue of *Morenci Copper Review*.

Al Fernandez, noted Clifton historian, tells that Jack Christie operated an old underground mine at that location.[2] The portal of the mine opened up onto the old highway south of where the long-gone town of Metcalf had been located. Many years after the mine was abandoned Phelps Dodge reopened it, extended the workings and created diamond drill stations for exploration drilling of the huge low-grade copper ore body of the Western Copper property.

The New England and Clifton Copper Company was formed in 1903 when the New England and Clifton Copper companies merged. The mines were located high in the mountains bordering the west side of the San Francisco River north of Clifton. The company also owned two underground mines in King Canyon: The Boulder and Antietam. North of their property was the Polaris Mine. The workings and dump can still be seen on a mountain towering above the San Francisco River valley.

A small underground operation, the "Black Cat Mine," was active after World War II. It was located on the Molinar claim above Garfield Gulch near Chase Creek, north of the site of the old town of Metcalf. The operator was Charlie Stevens. The ore mineral was chalcocite, a copper sulfide. Charlie had an agreement with Phelps Dodge stating they would buy his ore as direct smelter feed as long as it assayed 2% copper or better. Charlie mined the Black Cat during the 1950s and early 60s. He built a little house on the bank of Chase Creek and lived there with his wife, Grace, and their two sons, until his death. In recent years the copper deposits of the Garfield area once again became a focus of interest. This time as an exploration target of Phelps Dodge geologists.

The spirit of these hardy souls who staked the claims for these companies lives on into modern times. One of these early prospectors was W.D. Roper who staked many claims in the foothills of the Gila Mountains southwest of Morenci. After the sale of these claims to Phelps Dodge, Mr. Roper used part of his earnings from the sale to create Roper Lake, a popular recreation area south of Safford. Mr. Roper was then in his seventies. One would think he would have taken his money and retired, but his driving spirit kept him looking for the big strike of ore. Even after the sale, he could still be seen driving his old Dodge pickup with his helper riding in the bed of the truck, exploring the canyons and ridges. There is a story that his wife put half the money in a trust so he wouldn't spend it all on prospecting ventures.

---

[2] Al Fernandez, conversation with Ted Cogut, 1992.

This is the Coronado Trail in a photo taken by Howard Sherwood for the June 1952 issue of *The Prospector,* a Phelps Dodge Morenci Branch publication of the 1949-1952 period. The location is a short distance north of the Metcalf Mine. The photo description included the following: "During the past year the highway was paved, with the exception of a four mile stretch, as far up the Trail as Cherry Lodge. Extensive improvements have also been made at higher elevations including the relocation of the highway in the vicinity of Rose Peak (see map, page 6), by which some of the steepest and narrowest grades in the old road have been eliminated."

In July 1997, Grace, Charlie Stevens' widow, visited the site of the house they had lived in at the Black Cat Mine. Only the fireplace chimney remained. Nearby, a large boulder was found upon which had been inscribed: "1884 Stevens." *Photos by Don Lunt.*

19

Cal" Callicotte remembers the old town of Shannon...WCC

I met an old friend one day while stumbling around the grocery store trying to match a handful of coupons with countless brands of various and sundry items stacked high on the shelves. Glasgow M. "Cal" Callicotte was there pushing a grocery cart around. I see Cal occasionally since he retired about 20 years ago. He and I both worked in the Phelps Dodge engineering department for many years. Cal was the company architect and later became head of the Phelps Dodge rental department.

I first met Cal in 1951 when I started working for Phelps Dodge. Anyone who knows Cal is impressed with his stature. Tall — 6 foot 5 inches — raw-boned and slender, he made me think: "This must be what Honest Abe looked like — minus the beard!"

Cal is 86 years old. Thinking he would have memories that would help fill in our knowledge of the early days, I asked Cal about those times. We didn't have much time to talk because of the pressing business at hand (grocery shopping) so he just gave me a quick synopsis.[1]

He was born in 1910 in a little settlement that was small compared to the towns of Old Morenci or Metcalf. This remote settlement was located north of the Shannon Mine. High on the towering mountain, the inhabitants had a bird's eye view of Chase Creek Canyon. That was also north of the long-gone town of Metcalf, which was located where our present mine shops and truck garage are now.

Cal's birthplace was called "Metcalf-Shannon," frequently referred to as just "Shannon" and also, according to Al Fernandez, well known Clifton historian, as "Cerro Prieto" or "Black Hill." It was the company town of the Shannon Copper Company.[2]

---

[1] Conversation with Glasgow M. "Cal" Callicotte, 1996.

[2] Conversations with Al Fernandez, 1996-97. Al wrote about the old town of Shannon in one of his history columns for the *Copper Era*. He wrote that there were about 300 Spaniards living there and a Mrs. Gustablados, who ran its boarding house, would try to please her boarders by serving the kind of food they had known in the Old Country. She got these hard-to-get items from Al's Clifton store.

The dumps of the old workings of the Shannon underground mines are still visible at the top of Shannon Ridge. Cal's father worked for the Shannon Copper Company in the supply department. Cal still has the 1908 letter offering his dad the job!

The little settlement was extremely isolated. The only way there was by a very steep winding mountain trail that you traveled by foot or burro, or, if you were lucky and knew the operator, you might get to ride up on the Shannon incline in one of the empty ore cars that was climbing back to pick up another load of ore from the mine. Cal's dad rode his mule or the incline to get to and from home. That incline was extremely steep, rising almost straight up from the Shannon Railroad in Chase Creek Canyon to the Shannon Mine at the top of the mountain.

Considering how isolated the little settlement was, it is hard to believe there was a hospital in Metcalf-Shannon. But there was, and that's where Cal was born. His family lived there until 1913, when they moved to Clifton.

Some relics of the old town are still there as well as some of the framework and machinery of the Shannon incline. Ted Cogut and John Korolsky explored some of the remains of the incline as pictured and described in an article in *Morenci Copper Review.*[3] The site of the old town is now abandoned by everyone except for some big-horned sheep who occasionally visit there.

Cal has lived in Clifton ever since his family moved there in 1913. We thank him for sharing his memories with us. I know he has a lot more to tell!

---

[3] November 1994 issue, *Morenci Copper Review*. We were alerted to the need to document these artifacts while they still remained by Don Biles, PDMI property administrator, and Ross Arrington, PDMI safety inspector.

Jo and Glasgow "Cal" Callicotte (right) at a retirement party in the late 1970s. Warren Fenzi, facing camera, talks with Arnold Winkler, Morenci's head carpenter. *Courtesy Phelps Dodge Corporation.*

Close-up view of the large cable-guide wheels at the top of the Shannon incline. Partly buried by mine material deposited over the years, when this photo was taken they were still impressive reminders of what was once a technological wonder. *Photo by John Korolsky, 1994.*

Some of the old hoist machinery is still in good shape at the top of the Shannon incline. This cable reel, now secluded in a tunnel, raised and lowered ore cars between the top of the incline and the bottom of Chase Creek canyon in the early 1900s. *Photo by Ted Cogut, 1994.*

A mule has just pulled ten ore cars to the Shannon incline from the Shannon Mine. The large cable guide wheels shown in the photo on page 178 are pictured here in actual use (in the building at the top of the picture). *Courtesy Phelps Dodge Corporation.*

The Shannon incline, pictured here in about 1910, was said to be the steepest in the world. *Courtesy Roberta Troxell.*

20

The early Clifton-Morenci smelters — Rock House to Smelter Hill....TLC

When you think of the difficulties that had to be overcome and the very real hazards involved, it seems almost miraculous that copper metal was being produced in the Clifton-Morenci District as early as it was. Copper smelting in the District was preceded by quite a number of necessary events, none of which were anything like matter-of-fact. Some were nothing less than heroic.

The ores, first of all, had to be discovered by men who had the courage to venture into a war zone — the Southwest was then the exclusive domain of the Apaches and they would contest the European-American presence with unrivaled tenacity.

But still the European-Americans came. The first to actually discover the copper ores of the Morenci-Metcalf area were Robert and James Metcalfe. This was in 1870[1] though mineralization in the general area had been observed earlier, in 1863, by General James H. Carleton's California Volunteers. The nation was then torn by the Civil War and Carleton's troopers were in the area to secure the territory for the Union as well as to offer what protection they could from the Apache raids. Carleton relayed that mineral discovery to Salmon P. Chase, President Lincoln's Secretary of the Treasury. He believed that gold in abundance could be found in this area and that copper in "quantity enough to supply the world"[2] would be found nearby in western New Mexico. Probably because the Civil War had diverted attention, for a number of years, our later-to-be-famed mineralized zone lay in the backwaters of public consciousness; and when the Morenci-Metcalf ores finally were exposed to the full light of day, gold, as it turned out, would not

---

[1] Roberta Watt Troxell, *The History of Morenci, Arizona,* Master's thesis, University of Arizona, 1956, 5; and Patrick Hamilton, *The Resources of Arizona,* San Francisco: A.L. Bancroft & Company, 1884, 191.

[2] Library of Congress, letter dated September 20, 1863 from Brigadier General James H. Carleton, U.S. Volunteers, Major 6th Cavalry, to Salmon P. Chase, Secretary of the Treasury.

be found in abundance. Instead, it would be copper to supply the world that would be discovered there.

A Civil War veteran of the Confederacy who had previously been in the West as a "forty-niner" in California and an Indian agent in Oregon, Bob Metcalfe headed back in that direction not long after the war's end.[3] In time, he found himself in Silver City, New Mexico Territory. From there he traveled to a place that one day would be called "Joy's Camp" and even later "Morenci." There, Metcalfe saw stone hammers in a trench that were used by contemporary or ancestral Apaches. The hammers evidently had been used to break apart and expose the brightly colored oxides, the green and blue stones, which they used as pigments.[4]

There is no evidence that the Apaches had tried to smelt the colorful rocks that would later be recognized as copper ores. Smelting of those fabulous Morenci ores began later, in 1873, after the Lesinsky group built an adobe furnace at a location called "Stone House" (or "Rock House") in Chase Creek Canyon. This was just below the incline leading down from their Longfellow mine. In the same year, the smelter site was moved to Clifton. The fan for furnace blast was driven by water power coming from a dam that fed San Francisco River water to a ditch constructed along and parallel to the river. This mining venture was organized as the Francisco Mining Company.[5] In 1874, reorganized as the Longfellow Copper Company, a little over one ton of ore a day was being smelted in an adobe furnace in Clifton.

Henry Lesinsky, the principal owner of the Longfellow Copper Company, was a world traveler who had engaged in gold and silver mining in Australia and in Nevada before coming to a stop in his travels at the outpost settlement of Las Cruces, New Mexico Territory.[6] At Las Cruces, his uncle, Julius Freudenthal, and his brother, Charles Lesinsky, took him in as partners in their mercantile business. Henry later opened a store in Silver City and it was in that capacity, as a merchant providing supplies to miners, that he met Bob Metcalfe who came to him with ore samples and an offer to participate in a brand new mining venture in the Arizona Territory. A partnership

---

[3] "Reminiscences of the West," installment No. 40, *The Copper Era*. This autobiography of Robert Metcalfe was published in weekly installments in 1951-52.

[4] James Colquhoun, *The History of the Clifton-Morenci Mining District,* London, John Murray, 1924, 32.

[5] Norman Carmichael and John Kiddie, *Transactions of the American Institute of Mining and Metallurgical Engineers Vol. LXX, 1924,* "Development of Mine Transportation in Clifton-Morenci District," 829; also George M. Robison, *A Brief History of the Clifton-Morenci District,* The American Mining Congress Meeting of the Arizona Chapter, 1929, 2.

[6] James Colquhoun, *The Early History of the Clifton-Morenci District,* William Clowes and Sons, Limited, London and Beccles, c1920, 21-51. Colquhoun, himself an important figure in the local history, wrote a sensitive characterization of Lesinsky that portrayed his resourcefulness and courage in the face of adversity.

developed and Lesinsky, along with other partners, began smelting copper ore. Maud Metcalfe James, in a letter written to Roberta Watt Troxell, quoted an article written by her uncle, Charles Metcalfe Shannon, in which Shannon described the smelting work. The article was printed in the October 24, 1901 *Copper Era*:[7]

> Stone and mud ovens, after the Mexican style, were built for smelting the ore. The blast was supplied by an ordinary blacksmith's bellows worked by hand. This work, which proved very exhausting, was performed by relays of Mexicans, who worked in shifts of one hour each. In this way the first copper ever smelted in Arizona was produced in 1873. This is a point in the history of the Territory worth remembering. The pioneer ingot came from the Longfellow Mine. This was probably selected because its ore was easily smelted, and it could be reduced at a moderate expenditure of effort, being very high grade copper.

Henry Lesinsky's decision to enter the copper-mining business at that time involved taking a big risk. In attempting to successfully mine copper at that early date, he was moving into unproven ground. In an AIME publication of 1887, his operation would be reported as having been the oldest mining venture in Arizona,[8] but that claim really should have rested with the Ajo District. What was probably meant was that Morenci became the first to successfully overcome its difficulties sufficient to sustain a permanent operation. According to Ira Joralemon, mining at Ajo in those early times was carried on "in a casual and intermittent way," mainly due to the difficulties in shipping the ore to the smelter at Swansea in Wales.[9] In his recently written history of the Ajo district, Forrest Rickard provides a comprehensive account of mining at that camp.[10]

Even if Lesinsky could discover a way of smelting his particular ores locally, there was the problem of delivering the copper to market and of getting supplies for the camp. A wagon road had to be built to get the ore down-canyon from the mine to the smelter in Clifton, a distance of about five miles. And to get supplies into camp and copper out, another wagon road had to be opened from Clifton to Silver City, New Mexico, 100 miles away. From Silver City, the copper at first had to be *hauled on wagons 1200 miles to Kansas City.*[11] This distance was later reduced to *only 885 miles* when the

---

[7] Maud Metcalfe James letter, c1950s.

[8] Arthur F. Wendt, *Transactions of the American Institute of Mining Engineers, Vol. XV, 1887,* "The Copper Ores of the Southwest," 28.

[9] Ira B. Joralemon, *Transactions of the American Institute of Mining Engineers, Vol. XLIX (1914),* "The Ajo Mining District," 593. Joralemon notes that the ore at first had to be hauled by bull team 400 miles to San Diego and later to Yuma where it was floated down the Colorado River to the Gulf of California and there shipped to Swansea.

[10] Forrest R. Rickard, *Exploring, Mining, Leaching, and Concentrating of Copper Ores as related to the Development of Ajo, Arizona,* 1996, 3-8.

[11] Colquhoun, 6.

ox and mule teams began hauling their copper cargo to the railroad at La Junta, Colorado.[12] Most importantly, the ore body was by no means defined. It was quite possible there wouldn't be enough copper to make the venture profitable. And the war with the Indians who were protecting their homeland from the European-American intrusions was in full force. To make the picture complete, there were no amenities, no stores, no houses and no organized law structure to protect the miners from the rogue element that found good pickings in this lawless West. Quoting again from Charles Shannon: [13]

> The pioneer developers worked with varying success from 1873 to 1882, overcoming difficulties that would have discouraged any but the stoutest hearts. The country was frequently overrun by hostile Apaches, who attacked freighting teams, killed the men and ran off the stock. All supplies had to be transported with the teams from Kit Carson, Colorado, a distance of 800 miles and the copper produced was hauled over the same road.

Henry Lesinsky's adobe furnace was capable of smelting about one ton of ore a day. It required rebuilding after being used only about one week. Making copper this way was a tedious process. But a great improvement in smelter production was on the near horizon.

This paragraph from the 1887 *Transactions of the American Institute of Mining Engineers* described the smelting break-through:[14]

> In stopping up the holes burned by the corroding slag in the adobe lining of the small Mexican furnaces, a piece of copper was once accidentally used. It was noticed that instead of this copper melting, as was expected, it remained intact, the cooling action of the air being such that notwithstanding the smelting going on at the face of the bar of copper, the bar itself did not melt. This led to the building of one of these small furnaces with the copper-plate sides or walls; with the striking result that these plates, although reaching a red-heat, failed to melt, and answered the purpose of a refractory furnace lining. Long campaigns could thereafter be made without trouble. If a plate of copper did melt out, it was simply charged into the furnace and run out as copper bullion, and the plates themselves were cast on the spot.

This new improved smelting furnace was built in 1877, and with its use production rose to about 50 tons of copper per month. After a long struggle to make the mining venture a success, things were finally looking up for Henry Lesinsky and his partners, but the ensuing optimism was short-lived. The price of copper dropped from 25 to 15 cents per pound.

---

[12] Richard J. Hinton, *The Hand-Book to Arizona: Its Resources, History, Towns, Mines, Ruins and Scenery,* San Francisco, Payout, Upham & Co., New York, American News Co., 1878, 84.

[13] Maud Metcalfe James letter, quotes Charles Shannon October 24, 1901 article in the *Copper Era.*

[14] Transactions, 43.

Sentimental Journey

Of the 3,965 B-29s produced as a result of WWII, only a few remain. Of those, a very few remain that saw actual combat in the war.

One such is B-29-75-BW, serial number 44-70016. This airplane, produced at the Boeing Company plant at Wichita, Kansas, in March 1945, met its first crew at Kearney, Nebraska in April. There were only seven flight hours in the log book. This fresh, new replacement crew and 44-70016 were dispatched to the island of Guam, arriving on 28 April 1945, and were assigned to 458th Bomb Group Squadron, 330th Bomb Group (BG), 314th Bomb Wing (Very Heavy), 20th Air Force.

On Guam, the aircraft was officially designated as *K-40*, and appropriately marked. Within weeks the crew named their bird *Sentimental Journey* after a popular song of the time, and in the hopes of returning home soon. This name was painted on the left side of the nose. Also, in keeping with a 330th BG policy, an emblem consisting of a circle around a map of North America showing the United States and designating a specific American city was added to the right side of the nose. Several of the crewmen were from Philadelphia, PA; however, the name *City of Philadelphia* had already been taken. They settled upon *Quaker City* for the Group name.

K-40 and its Primary Crew flew a total of 24 combat credited missions while on Guam. The aircraft flew eight times with other crews and the Primary Crew flew one mission in another aircraft. K-40 was thus credited with 32 combat missions during WWII. The first mission of K-40 was against the airfield and installations at Matsuyama on the Island of Shikoku on 4 May 1945. One of K-40's last actual bombing missions in WWII was flown on 8 August 1945 when it sustained substantial battle damage and landed at Iwo Jima for repairs.

Two additional Show-of-Force missions were flown for combat credit, including being in the large B-29 gathering over Tokyo Bay and the Battleship *Missouri* when the Peace was signed on 2 September 1945. *Sentimental Journey* was flown back to Mather Army Air Field, CA in October 1945 by most of the Primary Crew who then returned to civilian life, unaware of the fate of 44-70016.

After the War

The record indicates that 44-70016 was placed in storage immediately after WWII and was reactivated in 1951. At this time it was assigned to the Air Defense Command's 4713th Radar Evaluation Electronics Countermeasures (REEC) Flight, based at Griffiss Air Force Base, NY. The 4713 REEC Flight named its aircraft after Disney's Snow White characters and 44-70016 was named *Dopey.* Its mission was to evaluate America's air defense system.

Rediscovery and Restoration

All B-29s were phased out of service with the Air Force in the continental USA with most being sent to the Military Aircraft Storage and Disposition Center (MASDC) at Davis-Monthan AFB, Tucson, AZ, to be cut up for scrap. Someone in MASDC noted that the B-29s were rapidly disappearing, resulting in a decision to offer one to the Pima Air & Space Museum just across the road from MASDC. The aircraft chosen was 44-70016. For years it sat in the desert heat with an assumed set of markings because the history was unknown. Then, through a set of unrelated but fateful circumstances in 1980, 44-70016 was rediscovered by the original crew. Within a short time, the crew rejoined K-40 at the Pima Air & Space Museum in the desert of Arizona.

The crew and PASM, working together, then set about to restore K-40 to its original WWII markings and configuration. In the mid 1980s the 330th Bomb Group Association was formed. They held their third Group reunion in Tucson with K-40 as the focal point. At that reunion the 330th voted to initiate a program to raise funds to put K-40 under some kind of shelter to prevent further damage from the elements.

Preserving the History

Meanwhile, PASM launched a program to contract a display building for its WWII airplane collection. The 330th Bomb Group and the Pima Air & Space Museum combined effort raised the funds and the hangar was constructed and dedicated in 1994.

The Arizona Aerospace Foundation is still accepting donations toward finishing the restoration of K-40 to it's WWII configuration.

Make checks payable to the AAF for the K-40 fund.

THE ARIZONA AEROSPACE FOUNDATION

6000 E. Valencia Road, Tucson, AZ 85706

Disaster loomed. In his *The Early History of the Clifton-Morenci District,* James Colquhoun described Henry Lesinsky's struggle to stay afloat:[15] At that time the Longfellow Copper Company's assets of $300,000 had been reduced by one-half. Lesinsky began to think they were struggling against fate itself and feared all their efforts were about to end in ruin. He greatly regretted what had become of his partner's investments and feared he would not live through the bad times.

But the cyclical nature of copper economics eventually brought a return of optimism and, by 1882, Lesinsky had realized enough from his venture to sell out to a syndicate of capitalists in Edinburgh, Scotland who organized the Lesinsky holdings and the Metcalfe holdings into the Arizona Copper Company. Henry Lesinsky, who built the first permanent house in Clifton, a fairly large single-story structure made of adobe that still stands, then moved to more cosmopolitan surroundings in New York City. He was among the first of a number of truly remarkable men who, right up to the present, gravitated to the mineral-rich mountains of Morenci, Arizona.

In 1875, three years after the Mining Act of 1872 made copper mining in the West a feasible enterprise, William Church of Colorado acquired control of a number of claims[16] located at a place first called "Camp Joy" and later "Morenci," in the Arizona Territory. The claims had belonged to the Detroit Copper Mining Company. Detroit Copper acquired them in 1872 from Colonel William Ryerson, Jim Bullard, Joe Yankie and John Swisshelm.[17]

The name *Detroit* Copper Mining came from the primary investors who were from the Detroit, Michigan area. Foremost among them was Captain Eber B. Ward, a millionaire.

We are accustomed to thinking that the early millionaires in Detroit all made their fortunes in the automobile industry, but Eber Ward became a millionaire when the automobile was still just a dream. He was the first Michigan man to become a millionaire from manufacturing enterprises. Ward's money came from such things as building ships for use on the Great Lakes, in the production of iron, from railroads and from glass manufacturing. The 1873 annual report of the Detroit Copper Mining Company indicates he held 3,658 shares in. By the time of the 1875 annual report, he held 5,160 shares.

---

[15] James Colquhoun, *The Early History of the Clifton-Morenci District,* London and Beccles, William Clowes and Sons, c1920, 45.
[16] Carmichael and Kiddie, 831.
[17] *Arizona Bulletin,* Solomonville, Arizona c1905.

The earliest subscribers in the Detroit Copper Mining Company and their shares as listed in the 1873 annual report were: D. Richardson, 1,829; J. Read, 1,225; J. Yankie, 1,225; W. Ryerson 1,225; W. Bronson, 458; W. Zabriskie, 457; C. Blodget, 457; M. Ward, 457; H. Ward, 457; H. Hodges, 457; W. Thompson, 457; D. Quirck, 457; A. Cook, 457; G. Gilbert, 457; R. Montgomery, 457; and C. Potter, 457. In the 1875 annual report the shares held by Read and Ryerson were up to 2, 857. An E. Pearce was also listed as holding 2, 857 shares, and an S. Burt held 1,200. Captain Miles Joy held 40 shares.

Some of the names of the investors and prospector/pioneers would live on as Arizona place names. Today's employees at the Morenci Mine will be familiar with Ryerson Canyon. One of the early underground mines was named the Yankie, there was also a mine and a concentrator named the West Yankie. And though the name Joy's Camp would be replaced by Morenci, there would, for some time, be a Joy shaft in the Arizona Copper Company underground mining operation. There are, also, the Swisshelm Mountains and a ghost gold-mining town called "Pearce."

The story of Phelps Dodge's introduction to William Church has been oft-repeated. It is a Morenci legend.

In his 1906 report, Dr. James Douglas, an early giant in the development of the Arizona copper industry, described a meeting he had in New York City with William Dodge of the firm Phelps Dodge & Company as the firm was then named. In the course of the meeting, William Church's visit was discussed. Church had just requested $30,000 from William Dodge and Dodge asked Douglas to examine Church's claims which were near the Longfellow Mine.[18]

In agreement with that request, Dr. Douglas met William Church in Lordsburg, New Mexico from which they traveled by stage road to Clifton and then up Chase Creek Canyon to Morenci.

After examining the Morenci site and also a site at Bisbee which William Dodge had also requested he do, Douglas rendered a favorable report on both properties. The result: on May 21, 1881 Phelps Dodge & Company acquired an interest in Church's Detroit Copper Mining Company and also acquired, in the next month, the Atlanta claim in Bisbee. This became a

[18] James Douglas, *Notes on the Development of Phelps, Dodge & Co.'s Copper and Railroad Interests,* 1906, Reprinted by Frontera House Press, P.O. Box 1011, Bisbee, AZ 85603, 12. The request has also been reported to have been for $50,000. Whether it was for $30,000 or $50,000, it was in either case a considerable sum for the time.

major turning point in the operations of Phelps Dodge which would, from that time forward, become increasingly engaged in copper mining.

Since there was no water readily available at the mine in Morenci, Church built his smelter on the bank of the San Francisco River about two miles downstream from Clifton. Due to summer-rain-caused delays in 1881, he was not able to complete construction and blow in his first furnace until the spring of 1882. This smelter was fraught with difficulty. Due to metallurgical problems, little copper was produced; and the Apaches, under the leadership of Geronimo, were on the warpath. They attacked the smelter from the other side of the river to divert attention from their plan of driving off the company's and contractor's livestock; and also attacked the mines, killing the teamsters who drove the ore wagons.

It was evident that some other way of operating had to be devised. To avoid the hazard of Apache raids on the ore wagons, the smelter could be moved to Morenci, near the mines. Or a railroad, whose engine would be of little use to the Apaches, could be built connecting the San Francisco River smelter with the mines. Church ultimately decided, on the basis of cost, to build a pipeline of about five miles in length which would provide San Francisco water at the mine in Morenci. This accomplished, in about 1883 he moved his smelter to Morenci.

About a year before Church moved his smelter to Morenci, Henry Lesinsky sold his property to the Arizona Copper Company, a Scotch company headquartered in Edinburgh, and Arizona Copper tore down his smelter, replacing it with one that was, according to Douglas, "far in excess of their needs."

After a number of desperate years when it looked as if the company would go under, Arizona Copper began to prosper under the innovative hand of James Colquhoun. Under his leadership, the company built a Bessemer plant in 1897 which smelted its concentrate at a record low cost.

The ACC smelter, located at the spot in Clifton where the Circle K store now stands, underwent a number of improvements over the years. One interesting innovation was the use of a mountain, the one that you can now view by looking directly behind the Circle K store, as part of the smelter's smoke stack.

In 1902, the Shannon Copper Company built a smelter near the residential area that would be known as "Shannon Hill," on the southeast outskirts of Clifton proper. Not much is known about this smelter. It was collocated with the Shannon concentrator. Both were sold along with the rest of the Shannon property to the Arizona Copper Company in 1919.

In 1913, ACC abandoned its smelter at the confluence of Chase Creek and the San Francisco River and began producing copper from a new facility built on the opposite side of the San Francisco that was downstream from the Shannon smelter and 2.2 miles from Clifton's populated area. The site would acquire the designation "Smelter Hill," by which it is still known.

This was a state-of-the-art smelter that was modern enough to not look greatly out of place even today. Its construction cost of $2,105,000 may not seem much by today's standards, but that amount gains significance when compared to wages that were then common. For example, these are samples of wages for eight hour's work that were paid to construction workers who built the smelter:

Laborer	2.00
Carpenter Helper	2.50
Boilermaker Helper	3.00
Labor Boss	4.00
Electrician	4.25
Boilermaker Layer-0ut	4.75
Machine Shop Boss	5.00
Pipe Fitter Boss	5.50
Brick Mason	6.50

To modern-day employees, these may seem to be small wages, but it was actually the going rate in the country. In 1914, the year after the smelter began production, Henry Ford would revolutionize Detroit's auto industry by more than doubling the wages of his employees, bringing them up to $5 a day.

Up to 500 men were employed in the smelter's construction which was described in detail by Chief Engineer E. Horton Jones in a paper presented at an AIME meeting in Salt Lake City in August 1914.[19]

Construction work at Smelter Hill began in February 1912 and was completed in February 1914 with construction far enough along for the facility to begin copper production in October 1913. The works contained three reverberatory furnaces that were 104 feet long by 27 feet wide and about 10 feet 6 inches to the crown of the arch. The furnace side walls were 2 feet 6 inches thick and the arch was 20 inches deep. Chief Engineer Jones reported that 363,638 bricks were used in the construction of the three reverberatories.

---

[19] *Transactions of the American Institute of Mining Engineers Vol. XLIX (1914)*, "Unit Construction Costs from the New Smelter of the Arizona Copper Co., Ltd.," E. Horton Jones, 3-155.

There were three 12-foot Great Falls converters, each having 28 tuyeres of 1 ½-inch diameter; two 40-ton cranes installed on the craneway; eight 7-hearth 21.5-foot-diameter Herreshoff roasters; two casting machines equipped with a steel cradle to receive a ladle of molten copper that was tipped by a 20 hp motor into a casting spoon of 1 ½-inch cast iron that poured into 39 conveyor-mounted molds, forming copper ingots with dimensions of 2 feet 4 inches by 1 foot 6 ¼ inches by 3 ¼ inches deep; and a 300-foot-high smokestack whose inside diameter was 22 feet at the top and 28 feet 8 inches at the bottom with an average wall thickness of about 24 inches. The foundation and bottom part of the stack had an unusual shape, being hexagonal instead of round.

In 1926, cracks appeared in the stack. The cracked bricks were replaced and 4.5-inch steel bands were placed around the stack, extending upward at 5-foot intervals up to within 45 feet of the top. In a paper published in 1929, I.J. Simcox, smelter superintendent, described the smelter and said this about the repair to the stack: "To date the cracks have not reopened and the stack appears to be in perfect condition."[20] Since the 1940s, when the smelter buildings were leveled, the stack had been standing alone beside the highway (now U.S. Highway 191). It was one of the first structures to capture your view as you approached Clifton from the southeast — a solitary monument to a once bustling copper-producing center. But on February 12, 1997 that last standing reminder of the days when men with Scotch accents roamed the area hills in search of ever more copper was blasted down.

That was to be the last of the Arizona Copper Company smelters. It passed over into Phelps Dodge's control when ACC sold out to PD in 1921 and was operated by PD until all operations were shut down in 1932 due to the Great Depression.

---

[20] I. J. Simcox, *The Clifton Smelter of the Phelps Dodge Corporation — The Mining Congress Journal*, The American Mining Congress, 1929.

This Detroit Copper Mining Company smelter was built in Morenci in 1884 and ran to 1920. The new No. 6 Arizona Copper Company concentrator can be seen under construction at the top of the hill and to the right Photo taken in January, 1906.
Courtesy Phelps Dodge Corporation.

In a photo taken October 10, 1912, we see a vastly different Clifton. This is at the confluence of Chase Creek and the San Francisco River, the bustling site of the Arizona Copper Company smelter. *Courtesy Phelps Dodge Corporation.*

Inside view of a smelter. This is thought to be the Arizona Copper Company's smelter in Clifton.

Courtesy Phelps Dodge Corporation.

Copper ingots are piled on the floor of the Arizona Copper Company smelter where Chase Creek joins the San Francisco River.
Courtesy Phelps Dodge Corporation.

Inside view of the Detroit Copper Company smelter in Morenci, early 1900s.
Courtesy Roberta Troxell.

Copper-producing operations inside the Arizona Copper Company smelter at Smelter Hill, southeast of Clifton proper. December 31, 1937.
Courtesy Phelps Dodge Corporation.

21

In the early 1900s, baseball was big-time in Morenci and Clifton...TLC

In an issue of *Morenci Copper Review* Bill Conger wrote about baseball in Morenci in the early 1900s. He had been researching the subject after receiving a request from Mr. and Mrs. Norman Crockett of Duncan who wanted to learn more about Mr. Crockett's father, Hite Crockett.[1]

Mr. Crockett gave Bill a photograph taken of his father in the early 1900s, wearing the uniform of the Morenci baseball team, and he mentioned that he thought his father had worked in the Detroit Copper Mining Company store. The picture, showing Hite Crockett with the rest of the Morenci team, all wearing the style of baseball uniform of that day and with **MORENCI** emblazoned across their chests, is a time-period classic.

The Crockett request led Bill into a bit of research. He (and I) wanted to know — in those early years — just how big *was* Morenci and Clifton baseball? As we were to learn, it was indeed big.

Later, when Bill was interviewing Frank Alvidrez and Frank Ruedas he learned that, as boys, the two had been scorekeepers for the Morenci team. They had traveled with the team when it played baseball in other towns in Arizona and New Mexico.[2]

In conformance with a policy of community support, the Detroit Copper Mining Company, a subsidiary of Phelps, Dodge & Company, also gave its support to the baseball team and did so in a number of important ways. Frank Alvidrez said the company would hire the players for some mining-related job and then give them an hour off with pay each day for baseball practice. The company also paid all their baseball-related expenses.

I was curious to know if Hite Crockett ever made it to the major leagues, so I made a phone call to my son-in-law in Maryland, Darrell Mach,

---

[1] May 1997 issue, *Morenci Copper Review*.

[2] Conversations with Frank Alvidrez and Frank Ruedas, 1997.

who has all the important books written about baseball. In *The Baseball Encyclopedia,* which lists statistics on all the players who ever played in the major leagues, he could find only one Crockett. It was not the one we were looking for. So Hite, though he may have been a good player, never made it to the major leagues. After all, how many players from any particular small town ever did? But the old-time Clifton-Morenci baseball story didn't end there. Bill had learned more about the Morenci-Clifton early baseball years from a book written by William Ryder Ridgway, noted Safford historian. Ridgway wrote about it in his Volume II of *Mount Graham Profiles.*[3] And in that chapter he also made note of two of the area's players who had indeed gone on to the major leagues — Chief Meyers and Bob Harmon.

Chief Meyers, the Indian boy from Riverside, California, had played baseball on the Clifton team in the years 1903-1904. His major league statistics are documented in *The Baseball Encyclopedia.*

John Tortes "Chief" Meyers was born July 29, 1880 in Riverside, California. He played nine years in the National League — seven for the New York Giants, from 1909 through 1915. In 1916 he played for the Brooklyn Dodgers and followed that with the year 1917 split between Brooklyn and the Boston Braves.[4] By the time he made it to the major leagues, he was 29, a little old for a ball player starting out. Would that hold him back — would his age keep him from achieving stardom? As we would learn, it was no handicap at all.

A catcher and a good hitter, Meyers had a .291 career batting average. He was not much of a long-ball hitter but did manage to hit six home runs in 1912, his best home-run year. But home runs don't tell us all of the story. Meyers' batting average that year of 1912 was .358, ***the league record for a catcher.***[5] In the previous year, 1911, he had a very respectable .332 average. In 1910, he had 62 runs-batted-in — his best year for that statistic. Chief Meyers played in four World Series — an unusual accomplishment for any player. In the 1911 through 1913 world series, he played for McGraw's famed New York Giants, following that up with World Series play for the Brooklyn Dodgers in 1916.

John Tortes "Chief" Meyers died July 25, 1971 in San Bernardino, California, a man who played baseball in Clifton and who achieved undying

---

[3] William Ryder Ridgway, *Mount Graham Profiles, Ryder Ridgway Collection,* Graham County Historical Society, 1988, 193-194.

[4] *The Baseball Encyclopedia,* Joseph L. Reichler, editor, MacMillan Publishing Co., New York, 1258.

[5] *The National League, a History,* revised 1995, Joel Zoss and John S. Bowman, Brompton Books Corporation, 23.

fame as a player in the National League. He had lived to be just a few days short of his 91st birthday.

Robert Green "Bob" Harmon, born October 15, 1887 in Liberal, Missouri, was another local baseball player who later became a major leaguer. Bob Harmon played for the Morenci team in 1907. After leaving Morenci, like Chief Meyers, Bob Harmon played nine years in the National League — five years for the St. Louis Cardinals, from 1909 through 1913, and four with the Pittsburgh Pirates, from 1914 through 1918, with the year 1917 off (probably in military service — the U.S. entered World War I that year). He was a pitcher.[6]

In 1911, Harmon's best year, he was a 23-game winner. In his second- and third-best years, 1912 and 1915, he won 18 and 16 games, respectively. For a pitcher, Harmon had an unusually good batting average, batting .255, .232 and .261 in 1909, 1912 and 1913 — his three best years at bat. His batting average and the fact that he hit a home run in 1914 may have accounted for the nickname the players gave him — "Hickory Bob." In the May 1997 issue of *Morenci Copper Review*, Bill Conger wrote about Bob Harmon:

> Harmon had heard about baseball in Clifton and Morenci and about good wages being paid by mining companies. He got a job in the Detroit Copper concentrator and worked his way up from $2.50 to $3 per day (darn good money in those days). His main love was baseball. Local betting was heavy whenever Clifton and Morenci clashed on the baseball diamond. Harmon didn't get to pitch much until a combined Morenci-Clifton squad played together and traveled to Albuquerque to enter a Labor Day tournament. Things got off to a bad start when our starting pitcher, a star from Deming who had come to Morenci to play baseball, allowed four runs in the first inning. Harmon was put in as a reliever and allowed no more runs, our boys still losing 4 to 3. Harmon did so well the coach left him in for the second game, which we won 4 to 3.[7]

Harmon went on to play semi-pro ball at Portland, then came back to play for the Morenci team sponsored by the Phelps Dodge mercantile store. He was to finally make the major leagues, playing for the Saint Louis Cardinals and Pittsburgh Pirates. After his retirement, he returned to Morenci in 1953 and was interviewed by William Ryder Ridgway.

Robert Green "Hickory Bob" Harmon died in Monroe, Louisiana on November 27, 1961 at the age of 74, a respected major league baseball player who knew Morenci well.

---

[6] *The Baseball Encyclopedia*, 1858.
[7] William Ryder Ridgway, 193-194.

A baseball team of the 1920s wears the Morenci uniform, well-known in its day. Left to right, front row: I. Spoon, 2d base; R. Gilliland, catcher; R. Boggle, shortstop; R. Templeton, manager; J. Heynegar, right field; L. Benitez, 2d base; G. Lovett, center field. Back row: P. Langridge, 3d base; J. Williams, pitcher; C.Tellez, pitcher; G. Smith, left field; F. Sandoval, catcher; and J. Baker, 1st base. *From Morenci Copper Review, May 1995.*

On this Morenci baseball team of about 1905, Hite Crockett, who played right field, is standing second from right. The names of eight other players on the team are known but, unfortunately, cannot be specifically identified in the photo. Nevertheless, they are: Williams, center field; E. Brubaker, pitcher; H. Brubaker, catcher; Hughes, first base; Notman, second base; McClellan, short stop; Holiday, third base; and Stanger, left field. *Morenci Copper Review, May 1997. Courtesy Mr. and Mrs. Norman Crockett.*

22

"Tex" Neel — a Morenci legend.... WCC

Few people achieve the status of "legend." One person who richly deserves the title is Larkin "Tex" Neel. His presence in Morenci helped shape the fortunes of the company and of the townspeople over many decades.[1]

Tex was born in 1878 on a farm in North Carolina. One day, Tex decided his father had whipped him once too often for playing mumbly-peg on Sunday, and he left home.[2] He was 14 years old and on his way to far-off Texas. Tex spent most of his early years in Texas working with horses. He was already called "Tex" when he arrived in Morenci in 1899 and was hired by Mine Superintendent Gordon McLean of the Detroit Copper Mining Company, the Phelps, Dodge & Co. mining subsidiary. Tex went to work in the boiler shop as a boilermaker helper.

Mules were used extensively in the underground mines of the early 1900s, and so it might have been expected that his work with horses in Texas would steer Tex toward the job of taking charge of the Detroit Copper Mining Company's mules.

In 1906, Tex got the job of surface foreman and kept it until his official retirement in 1947. His scope of operation covered a wide area, keeping him

---

[1] Telephone interview with Warren Fenzi, past president of Phelps Dodge Corporation and past general superintendent of the Morenci Branch , February 2, 1997. Warren Fenzi is the author of *An Oral History of Phelps Dodge*. He provided most of the information presented here about Larkin "Tex" Neel.

[2] Those of our readers for whom entertainment at home consists mainly of video games may not be familiar with mumbly-peg. When it began, we do not know, but it was commonly played by boys as late as the 1940s. It may date back to the invention of the jackknife. To play the game, two or three boys, usually, would sit on the lawn. With the long blade of the jackknife fully extended and the short blade extended only to the point of being perpendicular to it, the knife was tossed upward two or three feet. The way the knife impaled itself into the lawn would determine the winner. It did not seem to be the safest game in the world, but no one seemed to get hurt while playing it. Tex Neel himself was an example of the fact that boys of that time took on the responsibilities of adults at a much earlier age, and perhaps that was why they were obviously careful while playing. At a time when electronic games didn't yet exist even in the imagination, it was entertainment..

involved in almost every phase of underground mining and, later, open pit mining. His strong personality commanded respect from everyone who knew him.

Tex met and married a Morenci schoolteacher. When I first came to Morenci in the early 1950s, they lived in a beautiful house just south and over the ridge from the open pit mine. A main haul ramp passed near their house. To this day, that area is still called "Tex Neel Ramp." Very few people now working at Morenci know the origin of that name.

Phelps Dodge's horses were kept in a corral near the Neel's home. In the 1950s, we used the horses for surveying and geology work in the King-Metcalf area.

Tex and Mrs. Neel lived in that house a long time Like the man himself, the Tex Neel house became a fixture at the south edge of the mine. But, inevitably, even so strong a personality as Tex Neel had to cave in to the continuing search for new mining areas. The Neels finally left that house next to the mine to allow for mine expansion, and moved into a lovely home on Cedar Street in the Stargo area.[3] They had a commanding view of the ruins of the Arizona Copper Company No. 6 concentrator.

Tex's duties were not confined to a single occupational field. He was put in charge of security during the strikes of 1915 and 1917. During the 1915 strike, managers of the Detroit, Arizona and Shannon Copper companies, with due respect for the threat of violence, left town and set up camp in Duncan. Tex stayed in Morenci. With his gun at the ready, he took up a defensive position in a tunnel leading into the Humboldt Mine. When Governor Hunt ordered the militia into Morenci to regain law and order, Tex had the dubious honor of being one of those arrested by the militia. Years later, Tex went to visit the militia captain while the captain was stationed at the jail in Florence. But the captain didn't remember Tex. Being one of a number of persons that the captain probably had to deal with in that turbulent time, it is understandable that he wouldn't have remembered Tex.

Tex started lobbying with the Arizona legislature, a job he continued long after his official 1947 "retirement." He would go to every session with Folsom Moore from Bisbee. Both would stay in Phoenix for six-week sessions, talking to members of the house and senate. Tex had a lot of influence with the legislators. And he knew many of the Morenci managers. He was very influential with the managers, and they took his advice and counsel. Tex was also very close to the corporate officers including Phelps

[3] For those not familiar with the area, Stargo is located just a few miles south of the site of Old Morenci and about the same distance northwest of the present Morenci site.

Dodge Vice President and General Manager Harry Lavender, after whom the Bisbee open pit mine was named. Warren Fenzi, who provided most of this information, said: "They were both very strong guys and got a lot of respect."

Around 1950, with Tex acting as intermediary between the parties, a legal agreement was reached between the company and Gila River area ranchers concerning water wells in the vicinity of the tailing dams. He also released land to cattle ranchers, handling all the negotiations.

Tex was a legend in Morenci and still is. We young guys, who only knew him when he was quite old, would stand in awe in his presence. We all thought of him as the most powerful man in town. Even in those later years, Tex was respected by everyone for his toughness and fairness.

Taking time out for a picture in 1904: Bill Eldridge, Larkin "Tex" Neel, Jack Nelson. *The Prospector, September 1949.*

23

Early hydrometallurgical operations....TLC

The first production of cathode copper from the new solution extraction electrowinning (SX/EW) plant at Morenci occurred in September 1987. On November 10, 1987, in a ceremony that brought many visitors from the local area and others from as far away as Tokyo, we dedicated the plant. Included among the visitors was Arizona's Director of the Department of Environmental Quality, a new department that recently had taken over the environmental duties previously handled by the Arizona Department of Health Services.

It was a grand affair. The visitors were given tours of the facility, and the Morenci firefighters prepared their usual fabulous lunch. We made a brochure describing the process and a movie which was shown to the visitors at the Morenci theater. [1] In them we said the plant would produce 100 million pounds of copper annually, a projection that was met and exceeded in 1988 — its first full year of production — and would be quintupled with future plant expansions.

SX/EW was not new. Ranchers Exploration at its Bluebird Mine in Miami, Arizona had employed the process in 1968.[2] And, according to retired Phelps Dodge employees, Frank Ruedas and Pete Gomez, an early version of electrolytic copper production from copper-laden leaching solution had been tried at Morenci in the 1920s. That process would have lacked the particular reagents that later made the process a production success. In its modern configuration, at Phelps Dodge's Tyrone Branch in New Mexico an SX/EW plant had been built in 1984 and successfully operated by Tim Snider. He subsequently was transferred to Morenci to guide the construction and handle the production of the new Morenci plant.

---

[1]As an example of the cost-savings efforts brought on by prolonged low copper prices, I was assigned a number of jobs in addition to the environmental duties of the company's organizational chart. I was made the responsible culprit for the organization of the dedication ceremonies including production of the SX/EW brochure and supervision and editing of the movie script.

[2] *Pay Dirt* magazine, April 1968: "This plant will be the first to extract copper by solvent extraction."

Leaching, the first step in the SX/EW process, had put down its primary roots in the previous century, *and that had happened in Morenci.* This was a classic example of history repeating itself. Just as leaching, in the wake of the copper depression of the early 1980s, had revitalized the Morenci operation, it had also come to the area's economic rescue nearly a hundred years earlier. In 1893 James Colquhoun had saved Morenci's Scottish-headquartered Arizona Copper Company (ACC) from ruin with his introduction of the leaching process. According to his son, James Clifton Colquhoun, he began to leach tailings with sulfuric acid "made on the spot from inferior pyritic ore."

Like a drowning man who reaches out to grab anything that might possibly save him, the Arizona Copper Company was in desperate straits and was desperately looking for a life preserver. It had been losing money for years in its Morenci mining operation and the end of the company seemed both certain and near. Though one might think that at that dire moment they would have accepted any proposal that provided even the slightest hope of success, they did not embark on this leaching idea without exercising a certain caution. From their headquarters in Edinburgh, Scotland, they sent John Gill to Clifton-Morenci for an on-the-spot examination of the leaching procedure's prospects for success. John Gill's trip report, which immediately captures our interest with his highly detailed description of the leaching tests they conducted in Clifton, becomes truly fascinating when we consider that it was written in 1892, 105 years ago as of this writing. His report, printed in its entirety below, gives a good view of the leaching technology of the time:[3]

> The main object of my visit was to have practical tests made on as large a scale as appliances at Clifton would admit of, as to a suitable process for leaching our tailings and low grade ores. Among others, a process by Mr. Nicholas, of Liverpool, had to be tried. Mr. Nicholas accompanied me to Clifton to superintend the experiment of this process.
>
> Soon after our arrival we had experiments made with his process on practically all our low-grade ores and tailings; and at the same time similar experiments by means of dilute sulphuric acid were proceeded with. Excellent extractions were got by both processes. In the case of Metcalf and Coronado tailings the extraction was practically complete. We had a series of experiments with finely ground ores and tailings, and another series with unground tailings and with ores similarly crushed. In nearly every case the unground gave better extractions than the finely ground. Mr. Nicholas' process showed to some advantage in the extractions from the unground tailings, but when everything was considered, — facility in working, cost of plant, and net profit to be expected, — Mr. Colquhoun and I were quite satisfied that the sulphuric acid process is the best for our ores.

[3] Annual report of the Arizona Copper Company, Limited, 1892.

After coming to this decision, we had experiments made on quantities of 400 lbs. of tailings and ores at a time. Previous mechanical difficulties in agitating the tailings and dilute acid, and in procuring clear liquor by settling tanks and filtering appliances, were got over by circulating the dilute acid only, and using the body of tailings in the leaching tank as a filter bed.

This plan was successful in attaining the desired results, and with many advantages. Metcalf tailings gave an extraction of 98 per cent of the copper present. To the liquor thus obtained, which contained sulphate of copper in solution, was added scrap iron, and the whole copper in the liquor was got as metallic precipitate of copper.

The Metcalf and Coronado tailings and ores leach very readily. The Longfellow ores and tailings, owing to the presence of a larger proportion of earthy carbonates, and of iron and manganese, do not leach so well, and they require more acid.

I consider that my opinion has been confirmed by the expert who superintended the experiment near New York. He has allowed 1/2 ton of sulphuric acid to every ton of copper produced, in excess of the quantity used in his experiment, and 1/4 ton of iron per ton of copper beyond the quantity shown by practice to be required. He has also counted on acid being made from imported sulphur instead of from the Company's own ores. These allowances quite account for a difference of £7 per ton. No doubt variations will occur, and allowances should be made therefore.

The Board have now given instructions for the erection of an acid plant to manufacture 10 tons of acid per day, and a leaching plant to treat 100 tons of tailings daily. As the leaching process is quick in its operation, the extraction of the solution of sulphate of copper requiring only about three hours, the plant to be erected may be found to be quite capable of treating 200 tons of tailings daily, or it may be easily made to do so by very slight additions. The real limit will depend on the capacity of the acid works.

There is an abundance of fine leaching ores in the Metcalf and Coronado. They can be got on the surface from open-cast workings. Such ores in large quantities can be mined in the Metcalf, transported to Clifton, and crushed at a cost of about $1.50 per ton. In the ordinary case a proportion of high-grade concentrates would be taken there from, and the remainder would be tailings. The concentrates would pay for the cost of mining &c., and the tailings, costing nothing, would be available for leaching. Our experiments at Clifton showed that such ores could be treated by leaching without concentration, but after crushing and washing, and that after charging $1.50 per ton for mining &c., there would be a very considerable margin for profit, particularly if we manufacture acid from our ores.

This process of leaching, besides being profitable by itself, will lead to other advantages. It will increase our railway traffic, benefit our stores, and add to our rental as the owners of the town site of Clifton. If the Detroit Copper Company also adopt leaching, very material extra traffic must come over our railway.

Further, the process of leaching is preliminary to electrolysis. The solution of sulphate of copper procured by leaching is in an admirable condition for the application of the electric current. This would deposit the copper, recover part of the acid, and enable the use of scrap iron to be dispensed with. The proximity of the river will afford great facilities for the production of the electric current. An electrolytic plant cannot be erected on a large scale except at great cost, but it can be introduced on a small scale at little cost, and gradually extended without the cost being felt. The extensions consist mainly in adding to the number of the depositing troughs, and this can be done by keeping a few joiners at the work. The initial cost, however, of taking power from the river for a large electrolytic work would be considerable.

When the process was undertaken on the production scale, the copper solution derived from the leached concentrator tailings was combined with scrap iron in two revolving copper barrels. The end product, containing mostly copper, was fed to the smelter to produce copper ingots. Over a critical six-month period, this leaching system produced 518 tons of copper which otherwise would have been lost in the tailings. The new plant lowered the overall cost of mining so that more ore could be mined, making it possible to mine lower-grade ore bodies at a profit. Copper production increased 40% at a cost that was lower by 2 cents per pound. The increased profit paid the interest on the mortgage and paid for the leaching plant.[4]

James Colquhoun's leaching process saved ACC from disappearing. The company became profitable for the first time in its 11 years of existence. In 1895, the company paid a dividend to its shareholders — also for the first time.[5]

The Shannon Copper Company also adopted leaching. Bruce Kennedy, PDMI's chief engineer, has a text book printed in 1921 that provides a description of the Shannon leaching process which used a combination of sulfide and oxide ores.[6] In summary, the Shannon process consisted of the following:

> In order to eliminate the high cost of purchasing sulfuric acid to leach the oxide ores, the sulfides were roasted and contacted with barren liquor to set free sulfuric acid. The ore, coarsely crushed, was transferred to 25-foot diameter by 5-foot high tanks, leached with water by percolation, the copper solution run through a bed of raw oxidized ore and then through a bed of sufide ore or sulfide tailings. The resultant copper liquor was then contacted with scrap iron to form precipitates.

By the time that book was printed, Shannon had been sold to Arizona Copper which was, in turn, sold to Phelps Dodge.

---

[4] James Colquhoun, *The History of the Clifton-Morenci Mining District,* John Murray, London, 1924, 53.

[5] James Clifton Colquhoun, *James Colquhoun 1857-1954, the Man and His Work,* D. Leslie Perth, Scotland, 23.

[6] Leonard S. Austin, *The Metallurgy of the Common Metals,* "The Shannon Copper Company Process," John Wiley & Sons, Inc., 1921, 421.

At this location on the south side of Clifton, the Shannon Copper Company ran a leaching, concentrating and smelting operation. Shannon began operations in 1901 and sold its properties to the Arizona Copper Company in 1919.

Courtesy Phelps Dodge Corporation.

In the 1920s, these leaching vats in the town of Metcalf, which was by then becoming a ghost town, were used to make precipitate copper that was transported down the Coronado Railroad and smelted in the smelter at Smelter Hill, south of Clifton.
Courtesy Roberta Troxell.

24

The Longfellow Inn was quite a place.... WCC

There was a three-story building in the old town of Morenci near the head of Morenci Canyon that all old timers will remember as the Longfellow Inn. Located up against the canyon wall and just below "AC Hill," the Longfellow was a restaurant, bar, men's dormitory, apartment house and horse stable.

Old photographs indicate the Longfellow Inn was built around the year 1910. Al Fernandez, noted historian living in Clifton, is quite certain it was built by the Arizona Copper Company and named after the old underground Longfellow Mine.[1]

The restaurant, kitchen, banquet room, bar and a small room for private parties were located on the second floor. The top floor was the men's dormitory. The first floor was made up of four apartments and the unique feature of old horse stables. The stables occupied part of that first floor which could also be considered a basement as it extended back underground, beneath the hill, next to the building.

Marceline Marietti, who now lives in Ajo, lived in Morenci during the late 1920s and early 30s. She recalls that a Mr. and Mrs. Yee ran the Longfellow Inn. Their son, Jimmy, went on to become a surgeon in San Francisco.[2]

Marceline's first date was for a graduation dance held at the Inn in 1931. (She went on to graduate from the University of Arizona in 1935.)

She remembers a lawyer and another gentleman, Mr. Lennox and Mr. Ainson, who had offices on the bottom floor. She also says the building was originally a hotel and restaurant before the hotel part was changed to a men's dormitory.

---

[1] Conversation with Al Fernandez, 1997.
[2] Conversation with Marceline Marietti, 1997.

There is a young lady living in Morenci today who is very familiar with the old Longfellow Inn. Her name is Wanda Ruedas. Her mom and dad, Vina and Lyle Keenan, ran the Inn for many years, along with Ray and Tina Jacobs, Wanda's uncle and aunt. These are very familiar names to all of us who lived in Old Morenci.[3]

For many years, Wanda lived with her parents and her brother, Larry, in one of the first-floor Longfellow Inn apartments.

Wanda believes that Tina and Ray Jacobs took over operation of the Longfellow Inn during the middle 30s from the Yees. Ray soon added stairs to the front and back of the building to provide direct access to the restaurant and bar.

The horse stables under the building were still intact but no longer in use while Wanda was growing up. It was a favorite place for Wanda and her friends to play hide-and-seek.

Frank Alvidrez, long-time Clifton-Morenci resident, remembers when the stables were still in use. There was also a blacksmith shop associated with the stables, he said.[4]

Wanda actually started working as a waitress at the restaurant part when she was 12 years old and continued until she graduated from high school. She helped serve at Rotary and Lions Club meetings, company and Christmas parties — all held in the Longfellow banquet room. For large affairs, the Longfellow Inn staff also would help serve at the old Morenci Club.

Wanda's dad, Lyle, ran the kitchen and her mom, Vina, ran the "front end." Vina trained all the waitresses in all aspects of the job, including how to pick up glasses and plates. She allowed no gum chewing during working hours. Vina, Vina's sisters, Hilda and Tina, Wanda and Ilga Esparza were the regulars and were welcome familiar faces during meal times.

The Inn sold meal tickets for $30 per month. The prices of the meals were punched out of the tickets after each meal.

---

[3] Conversation with Wanda Ruedas, 1997.

[4] Conversation with Frank Alvidrez, 1997. Frank passed away in 1998, about six months after this interview. A chapter in volume II of this history is devoted to Frank's experiences with Phelps Dodge.

Wanda's parents left Morenci when the Longfellow Inn was closed. They presently own and operate the "Branding Iron" restaurant in Flagstaff.

I've enjoyed many happy times at the Longfellow Inn. On my first day in Morenci in 1951, I attended an American Institute of Mining Engineers meeting at the Longfellow Inn and became acquainted with many who later became close friends. The Inn was located across the parking lot from the engineering and geology buildings where I worked for many years.

Like many other notable buildings of Old Morenci, the Longfellow Inn finally had to fade into history. It was vacated and dismantled in 1966 to make room for the open pit mine expansion.

Thanks to Wanda Ruedas, Marceline Marietti, Al Fernandez and Frank Alvidrez for their helpful reminiscences of the old Longfellow Inn.

In this photo taken in the mid-1930s, we see the three-story Longfellow Inn. Built in about 1910, it was dismantled in 1966 to make room for the expanding Morenci open pit. *Morenci Copper Review, February, 1988.*

25

The many concentrators of early Clifton-Morenci —from the first in Arizona (1886) to the big "No. " 6 (1906)....WCC

By the middle 1880s, the grade of the ore of the two major companies then operating in the Morenci District, the Detroit Copper Mining Company and the Arizona Copper Company, had dropped to an average of about 6% copper content.[1] It was still very rich when compared to today's standards, but was too low grade to be successfully smelted to produce a high-grade final product.

The two companies had to find a way to upgrade the copper content before smelting. An ore concentrator was needed and, in 1886, the Detroit Company built the first copper concentrator in Arizona. It was located between Old Morenci and the little town of Longfellow that had sprung up above the Longfellow Mine. That same year the Arizona Company built its first concentrator.[2] This was located next to their smelter, in Clifton, where Chase Creek joins the San Francisco River.

In the concentrating process, the ore is crushed, then milled to a fine powder. The copper minerals are then separated from the pulverized waste rock. This is the concentrate, and it is ready to be smelted.

The two original Arizona and Detroit concentrators were practically identical. After the rock was crushed and milled, "jigs" were the concentrating devices. As the crushed ore was passed over the jig, it caught heavy particles containing copper and other metals, such as iron, gold and silver, in riffles. About 60% of the copper was recovered from the ore in this fashion. Each plant treated about 50 tons per day.

The tailing, or waste material, from the concentrators was still pretty high in copper content, so the companies later reprocessed it. The Detroit

[1] James Colquhoun, *The History of the Clifton-Morenci Mining District,* John Murray, London, 1924, 43.

[2] Colquhoun, 43,44. The Arizona Copper Company's concentrator began operating six months after the Detroit concentrator.

Company remined them for further concentrating and smelting. The Arizona Company treated their tails with acid to recover even more copper. This method was invented by James Colquhoun who was then general superintendent. The results pulled the company through a devastating depression of copper prices. In the early 1890s, copper sold for 9 1/2 cents per pound.

All the underground mines of the Detroit Company were located around Morenci. The Arizona Company owned mining properties near Metcalf, Morenci, the little town of Longfellow, and on Coronado Mountain.

The companies later would build other concentrators. The first five were designed to handle the "oxide" copper minerals of azurite, malachite, cuprite, chrysocolla, chalcanthite, brochantite and tenorite. By the 1890s, the oxide deposits were rapidly being depleted, and the two companies began to explore the sulphide ore deposits. The Humboldt ore body, which was located under Copper Mountain on the north side of Morenci, was a huge deposit of low-grade sulfide ore owned by the Arizona Copper Company.

In 1896, both companies built new concentrator plants for treatment of the sulfide ore, consisting mainly of the gray metallic mineral chalcocite. The yellowish chalcopyrite and bornite, which is also called "peacock copper" because of its iridescent copperish and purple colors, were present in smaller amounts.

The plants used two stages of concentration. The jig tails from the first stage were reground in mills followed by retreatment on vanners. Vanning was similar to panning for gold, where the tails were first agitated to get them into suspension. The heavy copper-bearing material was allowed to settle. The lighter waste material was decanted, leaving the copper-bearing minerals at the bottom. The result was a 20% copper concentrate and a tail (waste product) averaging 1.25% copper.[3] The tails themselves would be considered high grade ore by today's mining standards — the average grade of ore mined at Morenci today is about 0.6% copper.

Other methods of separating copper minerals from waste rock included sluice tables, Wilfley tables and settling tanks which utilized riffles, vibration, screening, rotation, leaching and panning (as in panning for gold).

The Arizona Copper Company built its first sulfide copper treatment concentrator in 1896 in Clifton next to its smelter at the mouth of Chase Creek.[4] This was its fourth concentrator. This had also been the site of its

---

[3] Edward H. Robie, "The Drift of Things to Come," *Mining and Metallurgy*, May 1942, 300.

[4] This was also *the* first porphyry concentrator. Colquhoun, 56.

first three concentrators. The first sulfide concentrating plant of the Detroit Copper Company was built overlooking Morenci Canyon. The plant turned out to be a failure; and, in 1897, William Church, president and general manager, sold the remaining part of his company to Phelps Dodge.

Unlike the experience of the Detroit Company, the Arizona Copper Company's concentrator successfully treated the sulfide copper ore. It was operated until 1914, using the two-stage concentration method described above.

The Detroit Company's original sulfide concentrator, which had been acquired by Phelps Dodge in 1897, was destroyed by fire in 1900. Phelps Dodge, still operating under the name of the Detroit Copper Company, built a new concentrator near the West Yankie mine shaft head frame, on the site of the original concentrator built in 1886. It was appropriately named the West Yankie concentrator. It overlooked Ryerson Canyon, which later became part of the original excavation for the open pit Morenci Mine. (Many years later, the tailing from the West Yankie concentrator, which was stored in Ryerson Canyon, proved troublesome to the electric shovels. The huge shovels would sink into the old tailing and get stuck).

In that same year of 1897, Arizona Copper Company built another sulfide concentrator in Clifton. It was an immediate success and was important in improving the company's profitability.[5]

In 1901, the Arizona Copper Company built a new concentrator near the old Longfellow Mine. This was in the area later called "Dispatch Hill," so named because it was the location of the dispatch tower for the trains operating in the open pit of the Morenci Mine. Like the West Yankie concentrator, it was situated on a point overlooking Ryerson Canyon. Called the No. 5 concentrator, it only ran for about four years and was abandoned when The Longfellow Mine closed.

After the turn of the century, the two most noteworthy concentrators were the West Yankie of the Detroit Copper Company, owned by Phelps Dodge, and the Arizona Copper Company's No. 6 concentrator. Built in 1906 in Morenci next to the Detroit Copper Company smelter, No. 6 could treat sulfide ore assaying 2% copper content. Both the West Yankie and the No. 6 made extensive use of belt conveyors, multiple-stage concentration and water recovery. Finer and finer grinding methods were developed. In 1907,

---

[5] Colquhoun, 71. Arizona Copper Company also had been running an oxide concentrator that was destroyed by fire on Christmas night of 1897.

a year after the No. 6 concentrator was built, a four-stage grinding system with concentration at each stage was employed at that concentrator.

In 1916, a new type of concentration was installed at the No. 6 concentrator. After the concentrate was crushed and milled to fine particles, water was added to produce a dilute pulp, which was subjected to agitation and frothing in cells. Chemicals were added to cause the copper-bearing minerals to stick to the bubbles and rise to the surface as froth. The froth was skimmed off as concentrate, and the material left at the bottom of the cell was waste material, or tailing. The process is called "flotation," and is the method used at the time of this writing in both the Morenci and Metcalf concentrators. In the No. 6, flotation was used in concert with jig, table and vanning systems to separate the copper-bearing minerals from the waste, or tailings.

To give the reader an idea of how huge these crushers can get, the primary crusher at the Morenci concentrator built in 1942 was a gyratory crusher, the whole assembly weighing nearly one million pounds. It was 57 feet from top to bottom and required nine railroad cars to haul all its parts across the country from the factory in Pennsylvania to Morenci. The crusher was retired after 50 years of service. During its lifetime, it crushed over 750 million tons of ore.

In 1917, the name "Detroit Copper Company" was laid to rest, changed to "Phelps Dodge Corporation, Morenci Branch." The company had been owned and operated by Phelps Dodge since 1897. In 1919, the Phelps Dodge "West Yankie" concentrator was shut down because of a severe depression in copper prices following the armistice ending World War I. Its production had increased from 400 tons per day in 1900 to 1,422 tons per day in 1919.

After Phelps Dodge acquired control of all properties in the district in 1921, all concentrating was done in the No. 6, which had been owned by the Arizona Copper Company. It operated until 1932, when all underground mining ceased. It was also used in 1939 and 1940 as a test mill for ore from the new open pit of the Morenci Mine. Its production capacity increased from 900 tons per day in 1906 to 5,100 tons per day in 1932. Up to that time, it was the largest, most productive concentrator that had been employed in the district.

In 1900, the Detroit Copper Mining Company, owned and operated by Phelps Dodge, built this concentrator to process sulfide copper ore. The shaft for the West Yankie Mine appears behind the concentrator.

Courtesy Phelps Dodge Corporation.

The Arizona Copper Company built this concentrator near the old Longfellow Mine in 1901. It remained in production for only a few years due to the abandonment of the Longfellow Mine. *Courtesy Phelps Dodge Corporation.*

A thickener tank for Arizona Copper Company's No. 6 concentrator is in the foreground and the town of Old Morenci is in the background. Circa 1925. *Courtesy Phelps Dodge Corporation.*

A side view of the Detroit Copper Mining Company concentrator appears at the left of the picture. The Liverpool shaft is seen in the middle and to the right. Above it, and slightly to the left, is the West Yankie shaft. In the extreme distance, near the top of the light colored mountain at the left of the picture, is the Arizona Copper No. 5 concentrator. The settlement that appears at the top of the mountain at the right of the picture is the little town of Longfellow. *Courtesy Phelps Dodge Corporation.*

26

Jack Cuthbertson: "It took comparatively little to keep everyone happy"....TLC

As we began to put the material together for this book, we were impressed with the number of people we met who had good stories to tell about life in the old local mining camps. Sadly, very few of them had thought of putting their memories down on paper. But Jack Cuthbertson was one who did. Jack wrote about life in Old Morenci of the 1920s and 30s in his interesting narrative titled *The Saturday Nickel.* He has been kind enough to let us pass on to you some of what he wrote.

Jack Cuthbertson retired from the Phelps Dodge Morenci Branch in 1982. Jack's forebears had come from the Cornwall tin-mining region of England. In the mid-1800s, many of these "Cousin Jacks," as the miners came to call them, went to work in the copper and iron mines of Upper Michigan. In addition to the metals they produced, metals that were sorely needed by a developing industrial country, the tasty pocket sandwich called the "pasty" that was filled with potato, beef, carrot, onion and rutabaga or turnip was a lasting contribution of the Upper Michigan miners.[1] This lunch-time treat is now on the menu in many restaurants in northern Michigan. After mining began in Morenci, Arizona Territory, the Cousin Jacks began coming here; and their mining expertise, gained through the toil of many generations, was put to good use in the development of the local mining companies.

I got to know Jack, his wife, Joyce, and their sons very well. Their two sons, George and Bill, are employed at PDMI. George works in the information services department and Bill works in the concentrator division after having been in the engineering department several years.

---

[1] Some say the pasty originated with the miners from the Scandinavian area, others with the Cousin Jacks. The term "Cousin Jack" supposedly originated when the English miners who, after being employed in a U.S. mine, would ask the foreman if he had a job for Cousin Jack back in England. When Cousin Jack arrived, he would send home for his "Cousin Jenny."

For a number of months, Jack and I served together on what was known as the "Non-ferrous Smelter Committee." Our job was to provide information for and advice in putting together a voluminous document required by the federal government which showed the financial impacts of the reconstruction that would be required to meet the new federal emission standards at the three Phelps Dodge smelters — at Douglas, Ajo and Morenci. The committee met in the Western General Office of the company back when that office was in Douglas.[2] I was on the committee to represent the environmental aspects and Jack, as chief accountant, was there to provide financial information. During the course of those trips to Douglas, I got to know Jack quite well.

Jack was born and raised in Old Morenci whose layout he described this way: "Morenci, I learned at an early age, like Rome, was built on seven hills."[3] One suspects this comparison originated with the Italian immigrants who had come to the area as businessmen or to work in the mines. (In 1913, the *Copper Era* reported Morenci-Clifton's population to be 15,000.)

The Detroit Copper Mining Company's center of operations lay in the midst of these hills, but it also had operations in the Fair Play area of the mining district. The other large mining firm, the Scottish-based Arizona Copper Company, centered its operations at Modoc Mountain, which was also called AC Hill, but had mines in the Metcalf and Coronado Mountain areas. ACC's Humboldt Mine was close to the Joy shaft that had been named after Captain Miles Joy who had surveyed the Detroit Copper Company's mining claims in the Morenci area in the early 1870s.

"Morenci had no paved sidewalks," Jack wrote. "There were, however, numerous trails around the hillsides leading to homes and different parts of town. Stairways of stone and cement or all cement as well as ramps were used for sharper changes in elevation." In regard to the steepness of the Morenci terrain, I have heard stories from old timers who said children, when playing in the steep yards, were tied to their homes with ropes to keep them from falling out of their yards!

"The area between the Morenci Hotel and the DC (Detroit Copper Mining Company) store," Jack wrote, "was paved with cement." Both were

---

[2] Nick Balich, who went on to become a Phelps Dodge vice president, chaired the committee. In 1982, the Western General Office was moved to the attractive Phelps Dodge Tower in Phoenix. The New York corporate offices moved to share the same building a few years later.

[3] Jack Cuthbertson, unpublished narrative titled *Saturday Nickel*, 1997. As for that title: "Children," wrote Jack, "were not given a stipulated allowance upgraded from time to time to keep pace with the cost of living.... Instead, if you had behaved reasonably well, had not committed too much mischief, had done your homework, had done your chores, you came to expect a nickel on Saturday morning that was yours to spend on whatever you chose, or, to save — if you were so minded."

impressive structures that had been built in 1901. A flight of cement stairs connected this area to the plaza, while a ramp provided access to the hotel from the plaza. Jack's description of the hotel area continued:

> There was also a walkway leading westward from the hotel along the side of the hill which was about fifteen to twenty feet above the tracks leading to the depot. This path was also below the red brick school (the Longfellow School) which was built next to the Morenci Hotel. At a spot opposite the Morenci Club it forked with the right fork ramping up to the playground. The left fork bridged the tracks and by a short flight of steps across the highway would put you at the entrance to the Morenci Club.

Jack described the construction of the Detroit Copper Mining Company store, the "DC store," that was said to be the finest department store in Arizona. It was located at the foot of D Hill, an area that was filled in order that a center of town could be created. The DC store was a four-story building made of limestone on its north wall and basement and otherwise of red granite quarried in Morenci Canyon and brought to the site via the Morenci Southern Railroad. Beside being an unusually large and fine store building, the DC store provided certain non-commercial attractions:

> Along the entire length of the DC store, on its south side at main entrance level, existed a wooden walkway complete with handrail. This was a splendid vantage point to observe who was coming into town, as it was just above the depot and was especially popular in the fall of the year, when the new teachers were arriving and the young bachelors could get a first-hand look at them. This was also a good spot to watch the twilight league games.

The Morenci Hotel was a picture of pure elegance and captivating architectural style. An issue of the *Bisbee Daily Review* provided a description of some of those prominent Old Morenci buildings that Jack mentioned including the Morenci Hotel which was described this way:[4]

> A generous provision for the entertainment of guests to the town of Morenci is the hotel erected by the Detroit Copper Company. It is in the mission style of architecture and construction, which is lavishly furnished throughout, one of the peculiar fascinations of the building being in the smoking room, where the fireplace is constructed out of cobble stones found in the canyon, selected for uniformity and perfectly smooth natural condition, so arranged in the mantel as to present a handsome appearance.

In his narrative, Jack Cuthbertson provides a good description of the interior of the old hotel. He notes there were actually two fireplaces, one in the second floor lobby and another in the third floor lobby, and that each had

---

[4] *Bisbee Daily Review*, "World's Fair Edition," 1904.

a fireplace with polished cherry-wood mantles. Jack also notes that the hotel, a three-story structure, was elegantly furnished and had a magnificent view down Morenci Canyon. It had the reputation of being the finest hotel between Denver and San Francisco. The entrance doors, transoms and lobby windows featured leaded glass. All the wood paneling and window casements were made of cherry wood. Herbert Hoover, while in his pre-presidential occupation of mining engineer, had stayed at the hotel.

After the hotel was dismantled in the 1970s to make room for the expansion of the open pit mine, one of the fireplace mantles, a handsome piece of craftsmanship made of cherry wood, was moved to the Greenlee County Historical Society Museum in Clifton. The museum was then in the Henry Lesinsky house, the oldest surviving home in Clifton. Water rose into the building during the 1983 flood and many of the museum's artifacts were lost but the Morenci Hotel mantle somehow escaped damage and was moved, in 1994, to the present museum on Chase Creek Street in Clifton where it can now be seen.

In 1974, when I arrived in Morenci, the engineering department would meet after work on Friday evenings at the Morenci Club in Old Morenci for refreshments and just to socialize. At that time, my environmental department, then called "emission control," was organized under the engineering department. I spent enjoyable evenings at that old Morenci Club, but these get-togethers did not last long as the building was soon scheduled for dismantling to make room for the advancing open pit mine.

The structure had been built at the site of the first Morenci store which was built by William Church in 1885. With Harry S. Van Gorder as its first manager, home grocery deliveries then were made by mule.

It was during one of these Friday evening meetings at the Morenci Club that I met Tony Enrico who later became chief design engineer. When he learned I was a meteorologist, Tony grasped the opportunity to tell me about the great snow of '67, a topic of discussion that we would have almost every time we met and one that I never tired of hearing. That momentous weather event in the history of Morenci is described in Volume II.

This was Jack Cuthbertson's description of the Morenci Club:

> Like the Morenci Hotel, it was of stucco construction. It should be mentioned that these two buildings were probably designed by the same architect — they were classics....they were functional as well as being pleasing to the eye. The basement of the club contained a bowling alley (two lanes), a billiard room and an exercise room containing mats, horses, horizontal bars and the like.

Prior to my arrival, the club had contained within its walls, as Jack explains, space leased for the Royal Theater, a Frank Martin business which, in addition to showing movies, sold ice cream, soft drinks and candy long before that became standard theater practice.

The only road with a name in Morenci was "Burro Alley." Jack's father told him that it was so named because "it was only wide enough to accommodate a burro train and if you happened to meet one you either jumped between two houses lining the alley or got knocked down. To my knowledge, there never was a sign post bearing this designation — you just got to know that was its name."

In the days when not many people drove automobiles, Jack tells us about Dr. Mann, a dentist, who practiced in Morenci and also in the Globe-Miami area. Dr. Mann had just purchased his first car from a dealership in Clifton. From Clifton, Jack noted, Dr. Mann had no trouble.

He negotiated the trip up to Morenci quite nicely. However, upon reaching his destination he was observed driving around and around a huge water tank which was located at the foot of Burro Alley. His garage was located in this vicinity, but when he finally decided to put the car in the garage it failed to stop and smashed through the back end, finally landing on the railroad tracks some 50 feet below. Miraculously, he was not injured and proceeded to explain that he had not been instructed on how to stop the thing, or if he had could not remember, but had grown tired of circling the tank, whereupon Wash McLean who had witnessed the entire farce asked, "Why didn't you just turn off the ignition?" To which the exasperated Dr. Mann rejoined, "Vell, vell, vhy didn't you tell me dat ven I vas circling the tank?"

Clifton-Morenci residents who sometimes think it is a bit bothersome to get caught behind a slow-moving truck on the climb up Highway 191 from the horseshoe curve should feel thankful that they do not have to travel the old road that brought travelers up that way in the early days. When I arrived in Morenci, though the road had been changed to its present configuration decades earlier, old time residents still remembered its frightful aspects. Jack describes it as winding along the contours above Chase Creek which was about 500 feet below. It was known to local residents as "Scotland's Curves" because the Arizona Copper Company mine superintendent, whose name was P.B. Scotland, failed to negotiate one of them.

Jack wrote about the recreational activities that children participated in, and there were quite a number. You would guess that, with Morenci's hills, coasting on homemade or purchased coaster wagons was quite popular. One game that present-day children would not know and that Jack said was a favorite, was called "kick the can." It was played with one of the food cans that your mother had emptied and you rescued from the trash. (Parents got away without having to pay as much for toys in those days!) I often wondered how widespread were the children's games of those days. In this case, both Bill Conger and I had played kick the can in our boyhood — Bill in New Jersey and I in Michigan. This is Jack's description of the game:

> Our favorite game was kick-the-can, a form of hide and seek where the can is placed in a circle and when it is kicked everyone scatters and hides, except the one designated *it* who must retrieve the can, place it in the circle, and then proceed to locate the other players and bring them home. Once they are all caught the other players must stay at home unless another player can sneak in and kick the can before *it* can spot him. If he is successful in kicking the can, everyone scatters and *it* must start all over again. This goes on until *it* is successful in rounding everyone up before anyone can kick the can. The one that was caught first then becomes *it*.

In the chapter on Tex Neel, we wrote about the game that boys played called "mumbly-peg" that Tex had played in North Carolina. Both Bill and I had played it, and I wondered if it had spread to the West. I asked Jack about it. "Oh yes," he said, "you played it with a jackknife and flipped it off your knee and elbow." What was flipped was an opened jackknife with its long blade fully extended and its short blade opened to the 90 degree position.[5] The scoring consisted of the way the knife impaled itself in the ground. Parents of today might think that game was a bit dangerous, but I never knew of any boy who was injured playing it.

There is much more in Jack's narrative that the reader is sure to enjoy. Perhaps we can persuade him to have it published. Before leaving this topic we want to add one more thought from Jack's narrative:

"When I began thinking about jotting down some of the memories of how Old Morenci looked, how people and business got around, what people did to amuse themselves, it came to mind that it took comparatively very little to keep everyone happy, or so it seemed to me."

---

[5] My brother-in-law, Robert Nordstrom of Albuquerque, provided further enlightenment. In the way he played it, the jacknife was flipped off your wrist, elbow, shoulder, chin and he thinks perhaps also off the finger and even the nose.

In this rare photo of Old Morenci taken in 1901, the large building is the DC store (Detroit Copper) which was built in March 1901. Notably absent are the Morenci Hotel (it would be built later in that same year), the Longfellow School ("red brick school") and the Morenci Club which would be built on the bare ground to the left of the DC store. *From the Merna Robison collection.*

Old Morenci in about 1905. The Morenci Club is at left. Immediately to its right is the Longfellow School and to its right is the Morenci Hotel.

From the F. Remington Barr collection.

Old Morenci in about 1915: 1. Longfellow School; 2. Morenci Hotel; 3. DC Store; 4. Morenci Club; 5. Phelps Dodge General Office; 6. Presbyterian Church; 7. Joy Shaft; 8. AC Store (later the Safety Office, Employment office and American Legion); 9. Catholic Church. *Courtesy Roberta Troxell.*

Many school children in Old Morenci learned their ABCs in the "red Brick school," formally named the "Longfellow School." The building to the right is the Morenci Hotel.

Old Morenci in the early 1920s. The Morenci Plaza is in the foreground. A horse and buggy and a Model T Ford are parked in front of the Detroit company store, and a row of cars of other makes are in the center of the scene. *Photos courtesy Roberta Troxell.*

Old Morenci in about 1915. The plant in the foreground is the Detroit Copper Mining Company smelter. To its right and slightly above is the Arizona Copper Company concentrator. *From the F. Remington Barr collection.*

Jack Cuthbertson, at extreme left, poses with the accounting department staff in 1965, *Morenci Copper Review, May 1991.* Photo identification:

1. Ted Bond, 2. Charlie Lafferty, 3. Charlie Wood, 4. Larry Rapier, 5. Jack Rapier, 6. Mary Virginia Mayne, 7. Roy Richardson, 8. Bob Bartee, 9. Juan Gonzalez, 10. Tom Ulery, 11. Bob Harrison, 12. Wayne Holiday, 13. Frank Ruedas, 14. Mary (Snooks) Hall, 15. Ann Tysoe (Madson), 16. Paul Romero, 17. Paul Hutchinson, 18. Jim Coniogue, 19. Tom Patterson, 20. Jack Cuthbertson, 21. Pete Valero, 22. Wannell Baker, 23. Chester Kiehne, 24. Jorean Ballard, 25. Jeanette Filleman, 26. Molly Bingham, 27. Margie Lowe, 28. Kitty Williams, 29. Ted Blair, and 30. Cy Gomez.

27

Emma Ruiz Pando lived in old Metcalf....WCC

In 1916 when Emma Ruiz Pando was only a few months old, her family moved to the mining town of Metcalf from its sister mining town of Old Morenci. The Ruiz's had originally come to the area from Chihuahua, Mexico. Emma lived in Metcalf for many years as her father made a living as an underground miner. Metcalf has been gone a long time, and Emma has since lived in New Mexico and now in Tucson. She has many recollections of the old town.

Though she had never lived there, the family previously had made their home in the tiny town of Coronado located at the Coronado Mine high in the mountains west of and overlooking Metcalf. Her older brother was born there. Coronado disappeared long before Metcalf did. Emma's home in Metcalf was located part way up a hill and was one of the last houses at the south end of town. She loved the old town and says, "Everyone had a view and didn't have to pay for it." The house was located near the old 3-B store (Bueno, Barato y Bonito ; Good, Cheap and Pretty).

All the mines around Metcalf, which were owned by the Arizona Copper Company, shut down in 1921 because of a severe depression of copper prices. After Phelps Dodge bought the Arizona Copper properties in 1921, the mines remained closed except for the Coronado Mine, which closed for the final time in 1923. Emma remembers that the miners who remained in Metcalf, including her father, continued to scratch out a living by working in the abandoned Metcalf mines, extracting whatever ore they could. One of Emma's saddest memories is that, after leaving Mexico, her father was never able to see his mother again. He had to continue working to support his family and could not take the time off.[1]

After 1923, the trains on the Coronado Railroad kept operating, hauling to the Clifton smelter the ore which the miners were able to salvage. Phelps Dodge had purchased that smelter along with all the other Arizona Copper Company holdings. Gone were the days of passenger trains after 1921, but Emma recalls freight trains hauling ore from Metcalf to Clifton well into the

[1] Conversation with Emma Ruiz Pando, 1998.

1930s. It was a big day to see the narrow gauge train roll into Metcalf for another load of copper ore.

Metcalf was on the Coronado Trail, the road connecting Clifton to Hannagan Meadow. Horses and mules had pretty well been replaced by cars by the time Emma was a schoolgirl but her family didn't have a car. Even then, Emma recalls, the Phelps Dodge Mercantile store continued to use trusty horses and mules to make home deliveries in Old Morenci. Roads had not yet been built to the Old Morenci homes. At Metcalf, there simply were no home deliveries. Before Arizona Copper Company sold its holdings in 1921 the PD Mercantile building in Metcalf had been the ACC store. Metcalf had been a bustling mining town and even had a few telephones during the 1930s, "maybe four or five."

In 1933 or '34, a fire started in the Metcalf PD Mercantile store. It destroyed the store, the Catholic Church, the Stevens home, the home of the Jessie Simmons family (PD Mercantile manager) and the whole center of town. Emma recalls: "I remember the fire was all around us. The whole center of town burned. I woke the family to get out of the house."

About the only buildings saved were the post office and the Metcalf Presbyterian Church. Emma's father kept the church from burning by spraying it with a garden hose. The post office was across the tracks from the church. The Farnsworths, who ran the post office, gave Emma's father the hose to spray the church after wetting down the post office. "My father was very brave. He saved the church," Emma recalls. Her most vivid memory of the old town was her father's bravery as he continued to spray water on the church while fire raged all around him. The Metcalf Presbyterian Church continued to hold worship services until the late 1930s when the entire town was abandoned and dismantled.

The beautiful Metcalf High School building was only used for grades one through six after the mines had shut down. Junior High and Senior High students then were bused to Morenci. Made of solid white bricks, the Metcalf school had shops, a chemistry lab, home economics room and lots of classrooms. There was a baseball field on the school grounds and a big school yard. When the school was torn down, much of its brick was used for construction in Old Morenci. Emma remembers her brother learned to fold the American flag when he was in the 4th grade at the Metcalf school. Emma is so sad that such a beautiful school building had such a short life. It was built in 1917 and was only used for three years as a high school before the depression shut down all the mines causing the disappearance of two major companies, Shannon and Arizona Copper.

Mr. Lutz was then Morenci and Metcalf school superintendent. Every Wednesday he and the music teacher would visit the Metcalf school. The visit was eagerly awaited because then the children would get to sing.

Every Halloween some boys would ring the bells at the Catholic and Presbyterian churches. Emma's father said that wasn't right — the bells were to be used to call people to worship, so he stayed at their church all Halloween night and guarded the bell. That church bell is now one of the three bells prominently displayed at Shepherd of the Hills Church in Morenci. The other two are from the original Presbyterian churches in Clifton and Old Morenci.

One day, Emma and some friends climbed up to the Coronado Mine when she still lived in Metcalf. She remembers looking down from the top of the dizzying height of the Coronado incline. "It was so steep! A horrible thing!" She remembered hearing about the 1913 disaster in which nine men were killed when the rail cars in which they were riding broke loose and plunged down the incline.

"We left Metcalf in late 1934," she says. "The town lingered for only a few more years after that." Phelps Dodge owned all the land. The houses were torn down either by the company or by the individual. The company would pay each family for the house. "My father tore our house down and hauled the lumber to Morenci. Our house had very pretty doors." The doors and their house's beautiful wood flooring came from a home in the old town of Shannon, high up the mountain and on the east side of Chase Creek. Her father had salvaged the doors and the flooring when Shannon was torn down.

Phelps Dodge moved all the graves in the Metcalf cemetery since that area would someday be included in an open pit mining operation. They were moved to an area near the Morenci Catholic cemetery. Emma is disappointed that the new location for the graveyard is now in such poor condition.

Emma graduated from Morenci High School in 1940 and later from the University of New Mexico in Albuquerque where she met her future husband who was "Presbyterian from head to toe all his life." During World War II, he served in the Army Air Corps. They were married in New Mexico and she taught in Santa Fe, mostly bilingual courses, for many years.

Emma Ruiz Pando now lives in Tucson where she has retired. In October 1998, she attended the 100th anniversary of the Shepherd of the Hills Church in Morenci which evolved from her church in Metcalf, and the original churches in Clifton and old Morenci. She spoke eloquently to the group about life in Metcalf and the Metcalf and early Morenci Presbyterian churches. The podium she used came from the Metcalf Presbyterian Church.

Metcalf in about 1918, looking northwest. Tracks of the Coronado Railroad appear near bottom of picture. The large light-colored building in the middle background is the famed Metcalf High School. Immediately to its front and at the mouth of Santa Rosa Canyon is the Shannon Railroad trestle.

Courtesy Phelps Dodge Corporation.

Emma Ruiz Pando

28

From the sledge hammers of the 1870s, crushers came a long way....WCC

From the beginning of mining history, it has been necessary to break ore into smaller pieces before trying to extract the metal from the rock, be it gold, silver, copper or some other metal. There are many methods of doing this, from the old-time miner wielding a hammer to present-day crushing technology.

In early-day mining in the Clifton-Morenci district, the ore was broken in the underground mines by blasting it, then working it over with sledge hammers to make it small enough to be treated in the crude smelters.

Concentrators weren't needed in early days because the ore was so high in grade it could be directly smelted. In 1886, the first concentrators were built in the district. Two concentrators were built, one by the Detroit Copper Mining Company and the other by the Arizona Copper Company.[1] The job of crushing the ore was moved from the underground mines to the concentrators. The first mechanical crushers in the district were "jaw" crushers and "roll."

Jaw crushers broke the rock by "chewing" it between two metal jaws, one movable, the other stationary. The jaws were spaced at a specified distance apart which determined the size to which the rock would be broken before passing through. Roll crushers used two rotating metal cylinders that were spaced far enough apart to allow rock pieces to pass between them after the rolls had squeezed and broken the rock to a certain size.

These two types of crushers were used in all the concentrators from 1886 to 1932 when underground mining came to an end. Over the years, the size of crushers increased dramatically along with the increase of the tons of ore from all the mines.

---

[1] James Colquhoun, *The History of the Clifton-Morenci Mining District*, John Murray, London, 1924, 43-44.

In the early 1900s, two new types of crushers called "gyratory" and "cone" were installed in the No. 6 concentrator in Morenci. These were used with rolls and jaw crushers in the several different stages required to break rock small enough to process in the concentrator.

Both the cone and gyratory crushers employ a large spindle-shaped cone which rotates and pivots unevenly, leaving an opening on one side. This allows a rock to fall in, then the cone rotates against the rock, forcing it into a hard wall, which breaks the rock into smaller pieces. The action is similar to grinding something to a powder using a mortar and pestle. An opening at the bottom allows the rock to fall through when the pieces are small enough. The large spindle which rotates is called the "main shaft assembly." Some of the differences between the cone and gyratory crushers are size, the way rock is fed into the crusher, the way the main shaft assembly is suspended, the shape of the hard surface against which the main shaft assembly rotates, and the way size adjustments are made.

There were two stages of ore crushing in the No. 6 concentrator located in Old Morenci, primary and secondary. In primary crushing, a large jaw crusher broke the rock to 8-inch maximum size. Then two cone crushers broke the rock down to 7/8 of an inch maximum size. Secondary crushing then was done by shorthead crushers and rolls, which reduced the rock to "7 mesh," meaning the rock would be small enough to pass through a screen with seven openings to the inch.

Spike Pingleton, who accomplished many innovative money-saving changes in mill repair during his career at Phelps Dodge Morenci, retired as concentrator repair foreman a few years ago. Spike went to work for Phelps Dodge in 1941 in the old No. 6 concentrator. The No. 6 had been the largest, most productive concentrator in the Morenci district since it was built in 1906 by the Arizona Copper Company and owned by Phelps Dodge after 1921.

Spike began working as a laborer at the primary jaw crusher. He recalls secondary shorthead crushers and rolls at the head of the ball mills. That primary jaw crusher was moved to a location near the new bedding plant to crush rock from the silica quarry for the smelter after the No. 6 was abandoned.[2] The No. 6 had to be shut down in 1932 when the Great Depression caused all mining operations to cease and remain so for five long years.

---

[2] Conversation with Spike Pingleton, 1996. After retiring, Spike joined several other mine retirees who provided highly informative and very well-received mine tours.

This is Arizona Copper Company's No. 6 concentrator. It was built in 1906 and ran until it was shut down by the Great Depression in 1932. Thereafter it was only used in 1939 and 1940 to test ores from the new Morenci open pit.

Courtesy Phelps Dodge Corporation.

Spike Pingleton, seated, and the other popular tour guides took time out for this picture with tour guide head Arlene Boling. Standing from left: Lee McLemore, Charlie Dunn, Arlene Boling, Dick "Blue" Nations, Walter "Pete" Brooks and Bob Draper. Julio Tavison, another popular tour guide, was not available for the photo shoot.

Morenci Copper Review, February 1994.

29

Why Morenci?....TLC

This was the first thing our niece said that morning when she awoke in our Morenci company house after having arrived from the Middle West during the dark of the night: "Why Morenci, Uncle Ted?" In her mind it was a fair question. That niece, Cheryl Dettman, was wondering why I would have settled in Morenci, a fairly remote mining camp, after spending a number of years in somewhat exotic foreign lands and having lived the suburban life in cosmopolitan areas of this country. If I had interpreted her question in that way I might have responded that, after all those other experiences, living in Morenci seemed to me to be comfortable indeed.

But I took her question to mean: why was the mining camp *named* Morenci? Since my first arrival several years before, it was a question I often pondered. After all, Morenci was not a common place name. It was, in fact, a rarity as place names go. Someone, I believed, had to have had a good reason for selecting it. The word "Morenci" seemed to have been of special significance to the person who chose it.

Shortly after arriving in Morenci, I had read in one historical account that the town had been named by William Church of the Detroit Copper Mining Company and that the name came from a copper mining district in Michigan. That seemed logical to me and I thought no more about it until, one day in 1979, I happened to be on vacation in Michigan and was traveling along in my daughter Willa's car on a highway in the extreme southern part of the state, only a half mile from the Ohio border, when a sign pointing the way to "Morenci" caught my attention. Of course, we dropped off that highway and headed for the town. My immediate thought was that this could not have been the Morenci, Michigan that Morenci, Arizona had been named after because this area, near the Ohio border, was a farming region and definitely not mining territory. It was *600 miles* from the nearest Michigan copper mining district, those famed Upper Peninsula copper mines.

We took a few minutes to explore the town. Like Morenci, Arizona, I found Morenci, Michigan to be a charming little town. I took a photo of its main street and, not having much time, we left.

After returning to our Morenci, I looked at a number of atlases and discovered those two were the only towns in the U.S. named Morenci. Intrigued, I called the mining department of Michigan Technological University which is located in Michigan's Upper Peninsula near the copper mining district and, in fact, was a school that had been heavily dedicated to mining. After talking to a person in that department who would have known, I learned there had never been a mining district in Michigan named Morenci. Interesting!

Some time later I borrowed a copy of Roberta "Bobby" Watt Troxell's well-researched and excellent *History of Morenci, Arizona* that she had written as a master's thesis while attending the University of Arizona.[1] She noted that William Church would have had no reason to use that name (Morenci). The author of an article in the May 1942 issue of *Mining and Metallurgy* wrote that our Morenci was named after that town in southern Michigan and further stated that "maybe it was the birthplace or home of Captain Ward."[2]

Captain Eber Brock Ward, a Detroit, Michigan multi-millionaire, had sent Captain Miles Joy to the site that would later be known as our Morenci to survey the claims of the Detroit Copper Mining Company. Miles Joy's place of operation at that fabulous copper ore body was then quite appropriately known as "Joy's Camp."

The DCM company had been incorporated in Detroit in 1872. In its first annual report, dated July 1st, 1873, Eber B. Ward was listed as owning 3,658 shares, by far the largest number of shares of any of its stockholders. Given that information, it was predictable, to me, that Ward would have been the one to give our Morenci its permanent name. But, if he did name it, why would he have chosen *Morenci*?

Before digging into Ward's background I decided to try to eliminate the still lingering possibility that the word "Morenci" might somehow be connected to William Church. There was no historical society in Morenci, Michigan, but from a query to the town's mayor I received a reply from Mr. Kyle Griffith, a teacher, who was kind enough to send me a copy of an old newspaper article written by Hoig L. Gray, retired editor of the Michigan paper *The Telegram,* who had then (probably in the late 1960s) visited our

---

1 Roberta Watt Troxell, *History of Morenci, Arizona,* Master's thesis, 1956, 23-24.
2 Edward H. Robie, "The Drift of Things," *Mining and Metallurgy,* May 1942, 301.

Morenci. Gray wrote that a 1960 revised edition of the 1935 *Arizona Place Names* had stated "that Church named the location after his home town in Michigan." Among other interesting things Gray also wrote: "Some of the Phelps Dodge history suggests that Morenci, Arizona was named by Church for a Michigan "mining" town. But the only thing mined in Morenci, Michigan was thousands of tons of yellow clay which went into brick for early houses and tile to drain lush farm lands."

Bobby Troxell had been kind enough to provide me with the correspondence she had received as she researched the same question. The obvious, Church's hometown, was eliminated almost immediately. William Church had been born in Quincy, Illinois and lived there until entering business life, first in Chicago, then in Colorado and later here in Arizona. This would be corroborated by information that I later received from the Denver Historical Society. There had been no known "Morenci" in Church's life before his arrival in Arizona. I tried to also eliminate Miles Joy. He, too, at least in my mind, might have been a candidate for naming our town.

This took me to the State of Michigan Historical Library in Lansing, Michigan. The Joys that I found in their genealogy books either were not Michigan people or they had lived at an earlier time. While there I also delved a bit into Eber Ward's background. I had read somewhere that he was the father of Princess Chimay. The library had a biography written about a Princess Chimay who appeared to have been a charming lady, a French woman who had been exceptionally active in France's politics, but had lived a century earlier than any descendant of Eber Ward. I was not to find anything about any other Princess Chimay, but the biography made me wonder if there could have been a French connection in the naming of our Morenci.

At about this time, my daughter, Willa, and her husband, Al Swartz, had become quite interested in my quest. They lived in Okemos, Michigan, a suburb of Lansing, and made their own search of the shelves of the Michigan Historical Library. They could find nothing on Miles Joy and quickly eliminated Morenci, Michigan as being Eber Ward's hometown. Ward had been born in Canada, had lived in Conneaut, Ohio and then moved to Detroit.

Al then went to the internet, searched for the word "Morenci," and received an interesting reply — in the Italian language — in which the word "morenci" appeared in the phrase: "di accumuli morenci." This, we were almost afraid to say, was finally going to be the answer. The answer to the naming of Morenci was going to be found, we hoped, in the translation of the description of a park called the National Park of Pollino in which that word "morenci" appeared. The park was located near the southern end of Italy. But was that really logical?

I called Charles Spezia, well-known mainstay of the Greenlee County Historical Society and a descendant of Italian immigrants. I wanted to know if there had been Italian immigrants in the Clifton-Morenci area in the very early years. Charles indicated that they would have been here then. Most of them had been businessmen but some also had been miners.

To further nail this down I contacted Lydia Anderson in the foreign language department of the University of Arizona. She was an interpreter of the Italian language. She translated the paragraph that son-in-law Al had received on the internet in which "di accumuli morenci" became "accumulated moraines." A disappointment, there were no moraines in the classical definition — a glacier-deposited soil — in Morenci, Arizona. We were hoping the phrase would translate into something like "big copper mountain!"

I then asked Ms. Anderson if there could be another interpretation of "accumoli morenci." I didn't expect her to go that far, but she immediately suggested that she would ask several Italians and soon responded by FAX with their answer. "You were absolutely right!" she wrote. "You can also consider that "accumoli morenci" could mean accumulations from the morenci (logically, the morenci mines)." She was trying to be very helpful, but now I'd have to find a mine in Italy named "Morenci," and it would have to have been in existence before our Morenci was named.

The State Department in Washington, I thought, would be a good place to check. I talked to a Mr. Gene Sweeney of the Western European Section of the State Department. He consulted an Italian friend but was able to find nothing more than to confirm that the word "morenci" meant moraine and that it probably should be spelled "morenici."

I then went to the internet myself, searched for a listing of Italian universities, and sent this query to the University of Trieste: "I am looking for an Italian translation of "Morenci." This is the name of a copper mine in Arizona, U.S.A. Does the word 'Morenci' have an Italian meaning that would relate to mining?" This answer came from someone in their exploration geophysics group: "Thanks for the reliance, but unfortunately I know nothing about 'Morenci.' It could be something related to some dark stuff, for example "moro" means someone with dark hairs...could it be "Morendi?" Back to the drawing board...

In my effort to make sure I had covered all bases, I decided to try again to see if there had been a Morenci mining district or a mine of that name in the copper country of Michigan's Upper Peninsula. I had, of course, near the beginning of the search, checked with the Michigan Technological University with a negative result. But another check might be in order. This time, I called the historical museum in Michigan's Upper Peninsula county of Houghton. The museum was located in Lake Linden, a town that had historical connections to copper mining. I did not say that I was calling from Arizona. A gentleman who gave every indication of knowing his business replied that there was no mine or mining district with that name in the UP and in conclusion, said: "Morenci? That's in Arizona!" To the drawing board one more time...

Not that I thought it would help much but I decided that maybe the thing to do next was to place more focus on Morenci, Michigan. Maybe, if we could see how that town got its name it might in some way lead us to the answer of the naming of our town.

I called the U.S. Department of Agriculture Soil Service district office that had responsibility for Morenci, Michigan. They were located in the southern Michigan town of Adrian. I explained the reason for my call and a very helpful lady looked up the soil type on a map in a booklet titled *Soil Survey Series 1947*. "You may be on to something," she said, and went on to explain that, although the soil in the town of Morenci itself was not in the moraine formation, moraine soil did surround the town and only at a short distance away. Okay, so maybe Morenci, Michigan's name was derived from moraine. But even that would not follow well unless Italians had settled in that area. At this point in what was beginning to resemble a detective story, my wife, Marie, and I made another trip to Michigan where our daughter, Willa, son-in-law Al and grandson Nicholas drove us to Morenci, Michigan.

Like Morenci, Arizona, Morenci, Michigan may not be big, but it is a decidedly friendly town. We happened to arrive at lunch time and stopped in at the Village Inn restaurant. This did not appear to be a part of that familiar chain of orange-painted restaurants. From all outward appearances it had the looks of those family owned restaurant's where the fare reminds one of home cooking down on the farm. We were not disappointed. I only paid $3.80 for a sizable and tasty lunch. The highest priced dinner menu item was pork chops — at $5.85! On our way to the town library, named the "Stair Library," I looked down the street at the large shade trees and thought this could be my kind of town!

At the library desk I met Assistant Librarian Joyce Woerner.[3] "We're from Morenci," I said, pausing long enough for the entire library staff to gain the full import of the statement — "Arizona." They had known about

[3] Joyce Woerner and her husband later visited in Arizona and hoped to tour the Morenci Mine. They were unfortunate enough to have their arrival coincide with another of those copper price downturns and tours were not then being conducted.

Morenci, Arizona, and in that hospitable Morenci way they couldn't do enough for us.

I explained that I was looking for a listing of some of the pioneer settlers of their region, with the hope of finding the name Joy or Ward; and they immediately produced some very old genealogy books that listed the names of residents of Lenawee County, their county. There were no Wards or Joys. In a cemetery listing they found one Joy, a woman who had died in her twenties; but there was no indication of any descendants. We also looked for any Italian family names and found none, so it didn't appear that Morenci, Michigan had been named for the moraine soil type. I asked if they knew how their Morenci got its name, and they copied for me a few pages from a book titled: *Our Journey in Time, Morenci, Michigan 1833-1976.*

The settlement had been established in 1833 and, like Morenci, Arizona, had undergone a name change. It was first called "Brighton," but its name was changed in 1838. Quoting from the book:

> However, it was discovered that another community of pioneers to the north in the territory had prior claim to the name. There has been some controversy as to how the name "Morenci" was selected, some say it was an Indian name but others believe that it was taken from the name Mount Morency and they changed the "y" to "i".

We discussed the fact that there was a county in the northern part of Michigan's Lower Peninsula named "Montmorency." It was vaguely familiar to us as that very sparsely populated area that my father-in-law, Walter Nordstrom, went to for his annual deer hunt in the 1930s and 40s. I didn't think it was the place that the two towns of Morenci could have been named after because it had been, at the time of the naming of both towns, densely forested and probably not yet organized as a county. Like Morenci, Michigan, Montmorency County was not a mining area. It was, in fact, some 250 air miles from the Michigan copper mining area. Aside from the annual fall influx of deer hunters, the area was well-known for the timber it had provided during logging's hay-day of the late 1800s and early 1900s. In addition to more local applications, its lumber had most likely been used in the building of the cities, towns and farmhouses of the largely tree-less prairie states.

We were given a grand tour of the newly expanded section of the library, an expansion that is going to double the library's size, making a very fine facility indeed. The library staff were justifiably proud of the fact that the entire cost of the expansion had been done through seed moneys that had been bequeathed to them and which they had wisely invested over a number of years, no government funds had been used. These were definitely the kind of people you would like to have as neighbors. After that pleasant tour, it was off to the office of the *Morenci Observer,* the town newspaper.

As would befit a small-town newspaperman, the editor was friendly and quick of wit. I particularly liked the motto beneath the *Observer's* masthead: *where time is killed humanely.* I told him of my quest — the search for the origin of Morenci. "I thought you guys in Arizona knew that," he said.

" No," I replied, "we've come here to get the answer from you." He then went on to say he had relatives in Canada, and he thought the name was French. You wouldn't want to make too much of that, but the thought crossed my mind that the French connection might not be far off. Eber Ward had been born in Canada; and, much earlier, when the Territory of Michigan had been settled, many Frenchmen from Canada had been the settlers.

Following, once again, the thought of leaving no stone unturned, I called the office of the county clerk in Montmorency County. The county seat was in Atlanta, Michigan. Perhaps, if we knew how that county got its name, we might also know how Morenci, Michigan, 250 miles south, got its name and that might lead us to how Morenci, Arizona, 1,900 miles distant from Morenci, Michigan, got its name.

Julie Morton answered my call. She said Montmorency County was organized as a county in 1881 though the region had previously been named. She read a few excerpts from a pamphlet on the county's history. That location, like the two towns of Morenci, had also had its name changed. It was originally named Cheonoquet for a Chippewa chief who had been a party to four Indian treaties, the first in 1807 and the last in 1837. The reader who remembers I am a meteorologist will understand why I thought the chief had a fine name, Cheonoquet meant "Big Cloud." Ms. Morton sent me the pamphlet.

At that point I thought I had exhausted all possibilities of nailing down the origin of the name Morenci, and initially I paid scant attention when I received the letter with the Montmorency County pamphlet. But upon reading the pamphlet I changed my mind completely.

The brief history of that county included a letter dated April 23, 1951 from a Carl Henry of Alpena, Michigan to Frank Weber of Atlanta, Michigan. In it, Mr. Henry wrote:

> It is uncertain whom the name Montmorency was intended to commemorate, and there does not seem to be anyone of that name of sufficient prominence in American or Michigan history to justify this action.....There was a Duke of Montmorency, High Admiral of France, who, in 1620, bought the lieutenant

generalship of Canada and a few years later sold it again without ever having set foot on this continent. There was also a deLaval-Montmorency or Montmorency-Laval, the first Roman Catholic Bishop of Canada, an energetic, faithful churchman, who made great efforts to prevent the giving of ardent spirits to the Indians, and who for many years during his bishopric, from 1658 to 1684, exerted a very powerful influence in New France....

Carl Henry was evidently a historian of some accomplishment who was responding to assertions made to Frank Weber regarding the county's name. He continued with this significant remark: "I have an early Michigan map lithographed in 1840 and the name of the county is there spelled Montmorenci (*this would have been the first documented spelling of morenci that we know of*) which supports your correspondents' insistence of French origin...."

I thought that would be all that I could possibly find until, one day, one of my other daughters, Leta Mach, picked up a brochure at the Ohio-Michigan border in which there was a street map of Mio, Michigan. She lost no time in informing me that the main street of Mio, Michigan was named "Morenci Avenue." Thinking about how that came about, I went to Mio which is the county seat of Oscoda County, and visited the county courthouse. Displayed on the wall was a street map, evidently the very first, of Mio. It was dated 1881 and clearly showed that the name of the town's main street was Morenci.

The nice folks at the courthouse recommended that I call Nelson Yoder, a local historian. Mr. Yoder did not know how the street got its name but recalled seeing an article on how Montmorency County got its name and remembered that it was taken from a prominent Canadian — this would have been Montmorency-Laval. Since that main street of Mio is actually the state highway leading south from Atlanta, the county seat of Montmorency County, only 34 miles away, it was logical to assume that the name had been carried south from Atlanta or north from Mio. Both counties, incidentally, were organized in that same year of 1881. Mr. Yoder thought the connection to the Canadian French would have been logical since lumbering had been an important enterprise in that area and a number of French Canadians would have been employed in that industry.

And now to place all the pieces of the puzzle into some semblance of order. It seems logical that the first Roman Catholic Bishop of Canada, a man "who exerted a very powerful influence in New France" would have been a significant figure to someone like Captain Eber Brock Ward who was born a Canadian. Although not raised in a religious family Ward obviously had developed an outlook similar to that of a person who held such beliefs and would have been sympathetic to the causes that the Bishop Montmorency-

Laval energetically espoused. According to his obituary printed in the *Detroit Free Press,*[4] Ward never used liquor (which would have put him in sympathy with the liquor-curbing efforts of Montmorency-Laval) and, just a week before his sudden death, had given $1,000 to a church — a vast sum of money in 1875.

It seems quite likely that the inspiration for naming the two towns of Morenci (Michigan and Arizona), the main street of Mio and the county of Montmorency goes back to a similar source — to someone who, like Captain Eber Brock Ward, was an immigrant from Canada and who could have had great respect for the bishop named Montmorency-Laval, an important figure in the development of Canada.

It is interesting to note that these Morenci place names came about not long after Canada as we know it had been organized under a confederation of Canadian provinces by the British North America Act of 1867, providing those Americans with ties to Canada with a rather natural desire to commemorate things Canadian.

With three place names in Michigan — the town of Morenci, Morenci Avenue in the town of Mio and Montmorency County — Morenci, Arizona's connection to Michigan seems strong. But who could have transferred that Michigan name to the mining site in far off Arizona Territory?

Those two men from Michigan — Captains Miles Joy and Eber Ward — emerge as primary candidates. Miles Joy had been sent from Michigan by Eber Ward to survey the Detroit Copper Mining Company mineral claims. The site of those claims was then called "Camp Joy." It seems most probable that Miles Joy would have been content to have the Arizona settlement continued in his name. But his boss, Eber Ward, who was by far the majority stockholder of Detroit Copper, would have been calling the shots. If he chose to provide a new name for the settlement, that would have been quickly done.

And so ends my little research effort regarding the name "Morenci" of Arizona. I conclude that the town was named by Captain Eber Brock Ward who was born a Canadian. Since Captain Ward died on January 2, 1875 and the site was called "Camp Joy" in 1872, the year in which Captain Miles Joy was at work surveying the claims held by Captain Ward, Morenci, Arizona would have been named by Captain Ward in 1872, 1873 or 1874.

---

[4] *Detroit Free Press,* January 3, 1875.

MORENCI
1892

-A- SMELTER
-B- OFFICE D.C.M. CO & P.O.
-C- READING ROOM BUILDING
-D- BOARDING HOUSE
-F- LODGING HOUSE
-G- SUPT'S OFFICE
-H- CAGE SHAFT
-J- JOY SHAFT A.C. CO
-K- SAW MILL
-L- SCHOOL HOUSE
-M- BLUE SHAFT
-N- DETROIT MINE A.C. CO
-O- RES. DR DAVIS & MR. FRAISSINET
-P- RES. MR SWEENY
-Q- COPPER MOUNTAIN MINE
-SS- SALOONS

The main street of Morenci, Michigan in 1979.
Morenci Copper Review, February 1990.

This Morenci Avenue sign names the main street of Mio, Michigan. It is 215 miles north of Morenci, Michigan. The building in the background is the Oscoda County courthouse. *Photo by Ted Cogut, 1998.*

INDEX
(*denotes picture or map)

MAPS